# IMPLICATIONS OF PARTITIONING AND TRANSMUTATION IN RADIOACTIVE WASTE MANAGEMENT

The following States are Members of the International Atomic Energy Agency:

AFGHANISTAN
ALBANIA
ALGERIA
ANGOLA
ARGENTINA
ARMENIA
AUSTRALIA
AUSTRIA
AZERBAIJAN
BANGLADESH
BELARUS
BELGIUM
BENIN
BOLIVIA
BOSNIA AND HERZEGOVINA
BOTSWANA
BRAZIL
BULGARIA
BURKINA FASO
CAMEROON
CANADA
CENTRAL AFRICAN REPUBLIC
CHILE
CHINA
COLOMBIA
COSTA RICA
CÔTE D'IVOIRE
CROATIA
CUBA
CYPRUS
CZECH REPUBLIC
DEMOCRATIC REPUBLIC OF THE CONGO
DENMARK
DOMINICAN REPUBLIC
ECUADOR
EGYPT
EL SALVADOR
ERITREA
ESTONIA
ETHIOPIA
FINLAND
FRANCE
GABON
GEORGIA
GERMANY
GHANA
GREECE
GUATEMALA
HAITI
HOLY SEE
HONDURAS
HUNGARY
ICELAND
INDIA
INDONESIA
IRAN, ISLAMIC REPUBLIC OF
IRAQ
IRELAND
ISRAEL
ITALY
JAMAICA
JAPAN
JORDAN
KAZAKHSTAN
KENYA
KOREA, REPUBLIC OF
KUWAIT
KYRGYZSTAN
LATVIA
LEBANON
LIBERIA
LIBYAN ARAB JAMAHIRIYA
LIECHTENSTEIN
LITHUANIA
LUXEMBOURG
MADAGASCAR
MALAYSIA
MALI
MALTA
MARSHALL ISLANDS
MAURITANIA
MAURITIUS
MEXICO
MONACO
MONGOLIA
MOROCCO
MYANMAR
NAMIBIA
NETHERLANDS
NEW ZEALAND
NICARAGUA
NIGER
NIGERIA
NORWAY
PAKISTAN
PANAMA
PARAGUAY
PERU
PHILIPPINES
POLAND
PORTUGAL
QATAR
REPUBLIC OF MOLDOVA
ROMANIA
RUSSIAN FEDERATION
SAUDI ARABIA
SENEGAL
SERBIA AND MONTENEGRO
SEYCHELLES
SIERRA LEONE
SINGAPORE
SLOVAKIA
SLOVENIA
SOUTH AFRICA
SPAIN
SRI LANKA
SUDAN
SWEDEN
SWITZERLAND
SYRIAN ARAB REPUBLIC
TAJIKISTAN
THAILAND
THE FORMER YUGOSLAV REPUBLIC OF MACEDONIA
TUNISIA
TURKEY
UGANDA
UKRAINE
UNITED ARAB EMIRATES
UNITED KINGDOM OF GREAT BRITAIN AND NORTHERN IRELAND
UNITED REPUBLIC OF TANZANIA
UNITED STATES OF AMERICA
URUGUAY
UZBEKISTAN
VENEZUELA
VIETNAM
YEMEN
ZAMBIA
ZIMBABWE

The Agency's Statute was approved on 23 October 1956 by the Conference on the Statute of the IAEA held at United Nations Headquarters, New York; it entered into force on 29 July 1957. The Headquarters of the Agency are situated in Vienna. Its principal objective is "to accelerate and enlarge the contribution of atomic energy to peace, health and prosperity throughout the world".



Printed by the IAEA in Austria
December 2004
STI/DOC/010/435

TECHNICAL REPORTS SERIES No. 435

# IMPLICATIONS OF PARTITIONING AND TRANSMUTATION IN RADIOACTIVE WASTE MANAGEMENT

INTERNATIONAL ATOMIC ENERGY AGENCY
VIENNA, 2004

Sales and Promotion Unit, Publishing Section
International Atomic Energy Agency
Wagramer Strasse 5
P.O. Box 100
A-1400 Vienna
Austria
fax: +43 1 2600 29302
tel.: +43 1 2600 22417
http://www.iaea.org/Publications/index.html



**IAEA Library Cataloguing in Publication Data**

Implications of partitioning and transmutation in radioactive waste management. — Vienna : International Atomic Energy Agency, 2004.

p. ; 24 cm. — (Technical reports series, ISSN 0074–1914 ; no. 435)
STI/DOC/010/435
ISBN 92–0–115104–7
Includes bibliographical references.

1. Radioactive wastes. 2. Hazardous wastes — Management. 3. Transmutation (Chemistry) 4. Nuclear fuel elements. 5. Nuclear non-proliferation. I. International Atomic Energy Agency. II. Technical reports series (International Atomic Energy Agency) ; 435.

IAEAL 04–00387

## FOREWORD

Almost all States with a nuclear power capability consider geological disposal as the end point for spent fuel declared as waste and also for the long lived radionuclides and high level waste resulting from the reprocessing of spent fuel. However, several States are considering or investigating partitioning and transmutation (P&T) as a potential complementary route in the management of the radioactive material resulting from nuclear power generation.

P&T has the potential to open new avenues for long term waste management and to reduce the radiological hazard (in terms of magnitude and duration), to weaken the decay heat evolution history (e.g. by eliminating long lived heat producing actinides) and to reduce the quantities of the fissile and/or fertile radionuclides that pose proliferation concerns.

Whereas only the major nuclear power States are potentially capable of developing a self-supported P&T activity, States with more modest programmes are studying the impact of P&T on their own waste management programmes and strategies.

Recognizing this, and taking into account the increased interest in advanced and innovative nuclear fuel cycles and reactor systems, the IAEA initiated in 2001 a programme dedicated to preparing a report analysing the current status of P&T. Potential options for implementing P&T and its potential impact on waste management programmes and strategies were evaluated from an international perspective.

The first draft report was prepared at a meeting from 15 to 19 October 2001 by four consultants: L.H. Baetslé (Belgium), M. Embid-Segura (Spain), J. Magill (Germany) and N. Rabotnov (Russian Federation). A status report on the subject was prepared by L.H. Baetslé. During a Technical Committee Meeting (TCM) held in September 2002, a draft document was discussed, revised and substantially extended by ten participants and representatives of the IAEA Departments of Safeguards and Nuclear Safety. After this meeting, the report was finalized at a meeting from 7 to 11 April 2003 by the same group of consultants, also including L. Stewart from the USA.

The IAEA wishes to express its appreciation to all those who took part in the preparation of this report. Particular acknowledgement is due to L.H. Baetslé, who chaired the TCM and put great effort into the completion and technical polishing of the report.

The IAEA officer responsible for this report was R. Burcl of the Division of Nuclear Fuel Cycle and Waste Technology.

*EDITORIAL NOTE*

# CONTENTS

# 1. INTRODUCTION

## 1.1. WORLD NUCLEAR ENERGY SITUATION

The worldwide electronuclear capacity is about 350 GW(e), which can be subdivided into three blocks, each 100–120 GW(e): the USA, the European Union and the rest of the world. For strategic evaluations, a block of 100 GW(e) is therefore a representative portion of the electronuclear output on the world scale. Two fuel cycle options have reached industrial maturity: the once through fuel cycle (OTC) and the reprocessing fuel cycle (RFC) with recycling of plutonium and some uranium. Worldwide some 10 500 t HM of spent fuel is discharged annually from nuclear power plants and is either stored or reprocessed. Currently, the industrial reprocessing capacity amounts to 3900 t HM/a, which means that only one third of the discharged spent fuel can be reprocessed. This situation led to the total inventory of stored spent fuel increasing to 130 000 t HM by 2000, while only 70 000 t HM has been reprocessed and transformed into high level waste (HLW) or used in light water reactor (LWR) mixed oxide (MOX) fuel. Only a small fraction of recovered uranium has been recycled, and constitutes, together with the depleted uranium inventory, an additional radiotoxic waste type, which has to be stored or disposed of safely. Owing to the relatively low price of uranium, this evolution is expected to continue for some decades. Increasing amounts of repository space will become necessary to cope with these high inventories of spent fuel. On a worldwide scale two repositories of the size of Yucca Mountain need to be licensed for spent fuel, and one of the same size for HLW [1].

In some States the safety case of an underground repository has been thoroughly evaluated. Adequate engineering designs have been found and sometimes accepted by the regulatory authorities. However, it has been difficult to obtain public acceptance of a repository installation. Whichever strategy is followed, a repository for radioactive waste will need to be established, whether direct disposal, reprocessing or partitioning and transmutation (P&T) is pursued.

This report addresses the potential impact of P&T on the long term disposal of nuclear waste and evaluates how realistic P&T scenarios can lead to a reduction in the time required for the waste to reach an acceptable activity level.

## 1.2. MOTIVATION FOR PARTITIONING AND TRANSMUTATION

In this seventh decade of nuclear power, the issue of waste disposal dominates public opinion. The basis of this is the perception of risk associated with a decision to dispose of nuclear waste in underground repositories for very long periods of time. Such decision making clearly involves risk; however, refraining from such decision making also involves risk. How can the energy needs of a nuclear State be covered if nuclear power plants are shut down? This matter requires a reasoned analysis, taking into account not only the 'pros' and 'cons' of the decision but also the consequences of alternative courses of action.

It has become fashionable to advocate the 'precautionary principle' when dealing with sensitive technological issues. In most cases, the underlying argument is negative: in dubio, abstine. In contrast, however, the precautionary principle does not imply making no decision or postponing a decision – application of the precautionary principle implies active investigation of alternative courses of action.

Large inventories of long lived nuclear, and particularly fissile, material constitute the current radioactive legacy. Continuation of nuclear energy generation implies increasing waste streams, which will need to be disposed of. In the event of the use of the OTC, repositories of the size of Yucca Mountain will need to be constructed every seven to ten years worldwide.

If nuclear energy expands and is operated for hundreds of years, recycling and reuse of fissile and fertile material becomes a necessity; elimination of residual long lived radiotoxic nuclides by P&T would reduce the radiotoxic inventory and burden on future generations.

It is in this context that we see the motivation for a research and development (R&D) programme on P&T. P&T techniques could contribute to reducing the radioactive inventory and its associated radiotoxicity by a factor of 100 or more and to reducing the time needed to approach the radioactivity levels of the raw materials: uranium and its equilibrium decay products originally used to produce energy.

Great scientific progress has been made in metallic barrier building and in mining technology. However, nobody can fully guarantee total confinement of radiotoxic materials in human-made structures beyond 10 000 years. The reduction of the solubility of long lived radionuclides in underground aquifers may be a first step in reducing their migration. Partitioning from nuclear waste streams and conditioning of long lived radionuclides into stable matrices may be a second step in this direction before disposal.

The economic implications of a P&T policy are not negligible, however, and the responsibility associated with this societal choice has to be taken by the present generation for an unknown number of future generations. The decision to reduce the radiotoxic inventory of nuclear material by transmutation is an important one that has to be taken by the responsible authorities.

The main goals of P&T are:

(a) A reduction of the hazard associated with spent fuel over the medium and long term (>300 years) by a significant reduction of the inventory of plutonium and minor actinides (MAs).
(b) A reduction of the time interval required to reach a reference level of radiotoxicity inventory by recycling transuranic elements (TRUs).
(c) A decrease of the spent fuel volume by separation of uranium to enable more efficient storage or disposal. This should result in an increase in the effective capacity of a final repository. However, this approach might require special handling of strontium and caesium after partitioning.

## 1.3. PURPOSE AND STRUCTURE OF THIS REPORT

This report concentrates on the radioactive waste aspects of P&T; the main purpose is to provide useful technical information for decision makers on the expected long term consequences of present day decisions on waste management. The fuel cycle and waste management technology necessary to implement the P&T option is described. The reactor development necessary to achieve the transmutation yield is beyond the scope of this report, but has been covered in other publications [2–5].

Sections 1 and 2 are addressed to decision makers in order to inform them of the coverage of the report and on the implications and consequences of introducing P&T in an advanced fuel cycle scenario. Section 3 provides information on the non-proliferation aspects of P&T, with special emphasis on neptunium and americium. Fuel cycles, considered in the context of P&T, are introduced and discussed in Section 4. The general nuclear situation in the world in 2000 is taken as a reference point for the evaluation of the fuel cycle facilities needed to implement an advanced fuel cycle strategy with P&T as a possible back end stage in sustainable nuclear energy development. Sections 5 and 6 provide the reader with brief information on P&T, which is necessary for a better understanding of this report. Sections 7–9 present additional information, conclusions and recommendations useful in the decision making process.

The long term fate of nuclear waste in natural conditions is compared in Annex I with the behaviour of natural analogues (e.g. the Oklo natural fission reactor) on geological time periods. The ability of the present generation to guarantee the persistence of waste disposal structures is based on confidence in human technology and is illustrated by the preservation of human-made historical monuments erected some 5000 years ago. Annex II is dedicated to recent developments in inert matrix fuel (IMF).

# 2. POTENTIAL IMPACT OF PARTITIONING AND TRANSMUTATION ON RADIOACTIVE WASTE MANAGEMENT

## 2.1. RADIOTOXICITY EVOLUTION, HAZARD AND RISK

### 2.1.1. Spent light water reactor fuel

The long term hazard of spent fuel and HLW is associated with actinides, particularly the TRUs, while the short and long term risks are due to the mobility of fission products in the geosphere and the possibility of their entering the biosphere. Radiotoxicity (defined for the purposes of this report as the activity or quantity of radionuclides in spent fuel or HLW multiplied by their effective dose coefficients accounting for radiation and tissue weighting factors by ingestion, inhalation and absorption) refers to the adverse biological effects on humans from radioactive material in spent fuel. The radiotoxicity evolution of spent fuel is very well known and depends on the type of fuel and the attained burnup.

As an example, Fig. 1 shows the radiotoxicity evolution of LWR ($UO_2$) fuel at a burnup of 50 GW·d/t HM [6, 7].

To investigate the effects of different P&T strategies on radiotoxicity reduction, three cases have been considered in addition to the open cycle. The resulting radiotoxicity curves are shown in Fig. 1, in which the cross-over point indicates the time at which the radiotoxicity of the waste reaches the reference level. The following observations can be made:

(a) The open cycle: the spent fuel is directly sent to long term storage with no P&T. It takes 130 000 years before the radiotoxicity reaches the reference level.

(b) Full multiple recycling of plutonium as well as of americium and curium with high overall efficiency of P&T processes (99.5% for plutonium and

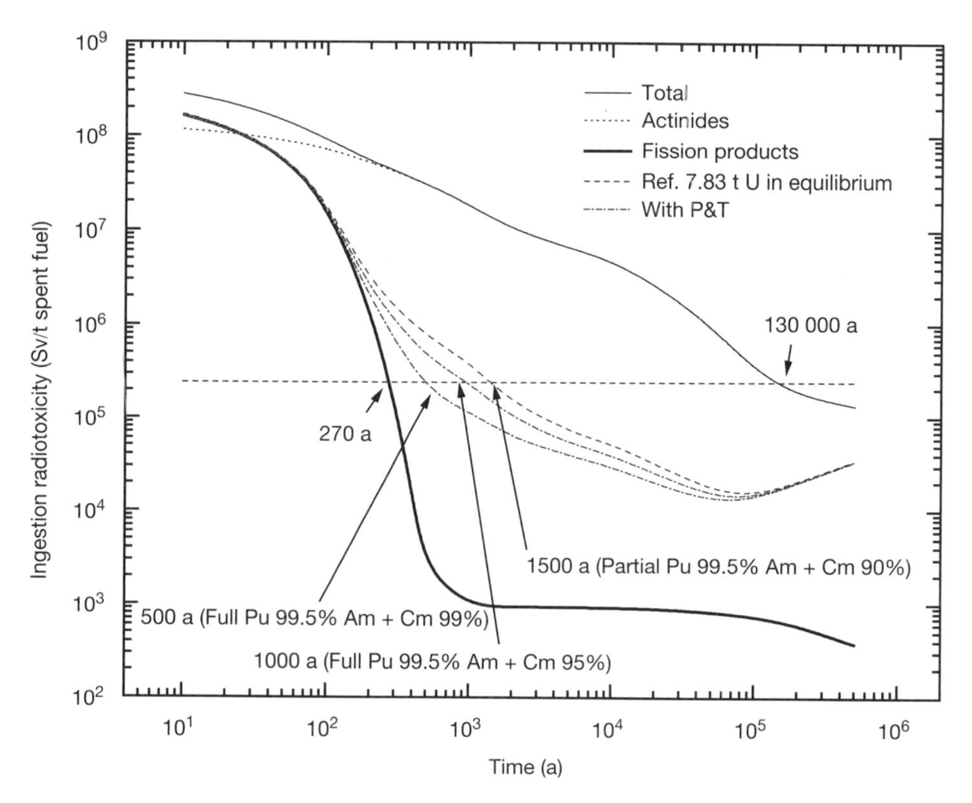


*FIG. 1. Ingestion radiotoxicity of 1 t of spent nuclear fuel.*

99% for americium and curium). The cross-over point is 500 years. If the curium is left in the waste, this time is extended to 1000 years.

(c) Full multiple recycling of plutonium as well as americium and curium, with less overall efficiency of P&T processes (99.5% for plutonium and 95% for americium and curium). The cross-over point is 1000 years.

(d) Partial multiple recycling: multiple recycling of the plutonium (99.5% P&T efficiency) and one single recycling of the americium and curium. In this case the americium and curium are transmuted in targets in a fast reactor (FR), with a 90% P&T overall efficiency foreseen. Thus the cross-over point is around 1500 years. In this strategy we can also consider leaving the curium in the waste; in this event 3000 years is required.

Based on these results, it can be concluded that P&T can help to reduce the time during which nuclear waste should be isolated from the biosphere from 130 000 years to between 500 and 1500 years. The fission products radiotoxicity curve gives the theoretical limit to the total radiotoxicity reduction in the event that all the actinides are partitioned and transmuted (i.e. no losses). This time is about 270 years.

During the first years the fission products $^{137}Cs$–(Ba) and $^{90}Sr$–(Y) determine the radiotoxicity and heat emission. From several decades to a period of about 250 000 years the plutonium and growing americium isotopes are the main contributors to the radiotoxicity, for the reference level of 7.83 t of uranium in equilibrium. Beyond 250 000 years $^{237}Np$ emerges, together with the progeny of uranium.

It is also of interest to see how the main components contribute to the total ingestion radiotoxicity. This is shown in Fig. 2, in which the results [6] are grouped according to the chemical elements present in spent fuel after six years of cooling. Note that 'U' refers to the sum of all uranium isotopes after six years of cooling. At later times the uranium isotopes decay to other chemical elements, which are also accounted for in the U curve. The advantage of grouping in this manner is that it is easier to see the effects of partitioning.

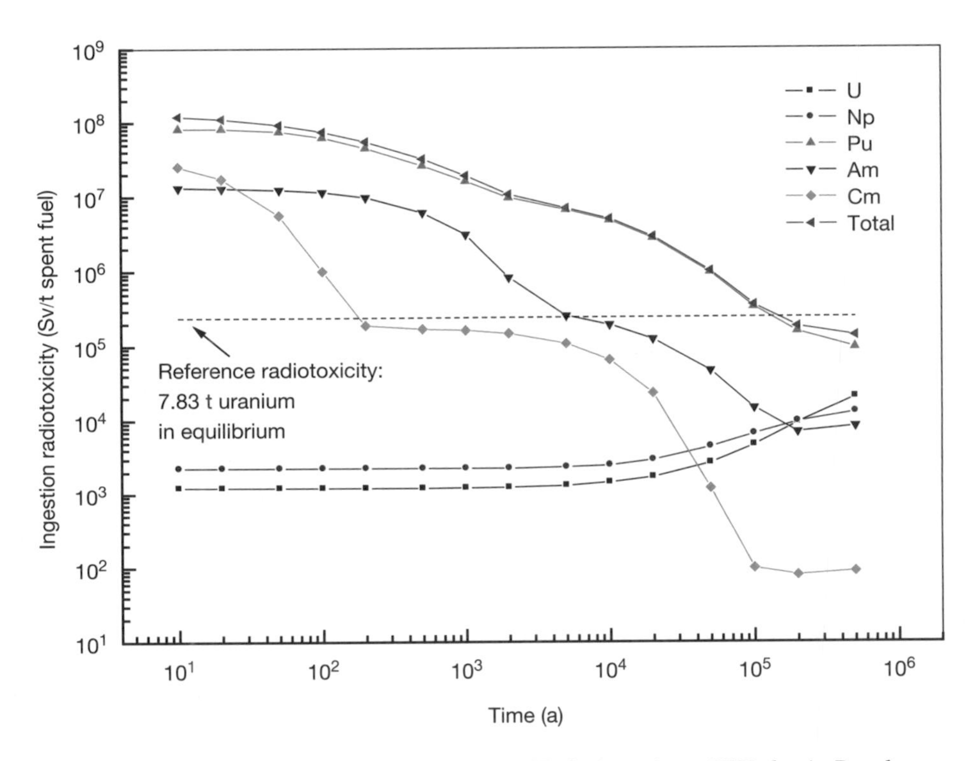


*FIG. 2. Actinide mass inventories in spent PWR fuel vs. time (ITU data). Results are grouped by element present after six years of cooling.*

As can be seen from Fig. 2, the total ingestion radiotoxicity arises from the plutonium isotopes. At around 200 years, the plutonium radiotoxicity is $4.3 \times 10^7$ Sv (i.e. a factor of 180 ($4.3 \times 10^7/2.4 \times 10^5$) above the reference value indicated by the dashed line). Conversely, to reduce the waste toxicity to reference levels on this timescale, the effective plutonium removal efficiency needs to be 0.994 (i.e. 1 – 1/180). Similarly, at 200 years the americium radiotoxicity is $9.4 \times 10^6$ Sv (i.e. a factor of 40 higher than the reference level). Hence an effective removal efficiency needs to be approximately 0.975 (1 – 1/40). The neptunium and uranium curves fall far below the reference levels.

For curium, it can be seen that the radiotoxicity at around 200 years is actually below the reference level, so it may not be necessary to remove it from the waste. However, this depends on the P&T strategy followed for the other waste streams. The curium radiotoxicity follows from 200 to 5000 years a plateau around the value of $10^5$ Sv. If the separation and transmutation of the plutonium and americium is efficient enough to significantly reduce the total radiotoxicity of the waste, then the curium could be left in the waste (its own radiotoxicity is then below the reference level).

Another approach to the evaluation of radiotoxicity evolution is presented in Ref. [8], in which a procedure is proposed for the calculation of radiation equivalence of waste and feed material in a fuel cycle with transmutation.

### 2.1.2. Spent light water reactor MOX fuel

The recycling of plutonium into LWR MOX fuel generates a much more radiotoxic spent fuel type, which up to now has remained in interim storage awaiting either disposal or recycling if a second era of FR MOX use becomes an economic reality within the next few decades. Since the decay heat of this fuel is much higher than that of LWR $UO_2$, the cooling period before recycling must be prolonged by a few hundred years. The radiotoxicity of this fuel follows the cumulated decay curves of $^{244}$Cm, $^{238}$Pu, $^{241}$Am, $^{239}$Pu, $^{240}$Pu and $^{242}$Pu. The other alpha nuclides (e.g. $^{243}$Am, $^{245}$Cm and $^{237}$Np) are in this context negligible.

### 2.1.3. Spent fast reactor MOX fuel

The alpha radiotoxicity of FR MOX fuel is approximately three times higher than that of LWR MOX, taking into account the increased plutonium concentration (~8–22%) in the fuel of a number of FRs that have been operating over several decades (DFR, PFR, PHENIX, SUPERPHENIX), but the radiotoxicity expressed per unit of energy produced is nearly the same.

The radiotoxicity of conceptual fuel of fast burner reactors (FbuRs) (CAPRA) and accelerator driven transmutation systems is undoubtedly an order of magnitude higher (up to 42% plutonium and a burnup of 210 GW·d/t HM).

### 2.1.4. Radiotoxicity of fission products

In the context of radiological risk reduction (concerning a deep geologic repository), the water soluble fission products $^{129}I$, $^{137}Cs$, $^{135}Cs$, $^{79}Se$ and $^{126}Sn$ are the most important radionuclides, due to a combination of toxicity, half-life and concentration.

The most toxic of the fission products is $^{129}I$, which has a specific toxicity (Sv/Bq) similar to that of the actinides. Caesium-137, one of the main fission products in HLW, has a toxicity that is ten times lower. The other soluble fission products are less toxic by two or three orders of magnitude compared with the actinides. Some fission products (e.g. $^{99}Tc$ and $^{107}Pd$) have long half-lives but are only slightly soluble in chemical reducing media representative of deep geologic repository conditions. The solubility of $^{99}Tc$ (as $TcO_4^-$) in oxidizing conditions (e.g. Yucca Mountain) is much higher.

### 2.1.5. Advanced conditioning of minor actinides

An advanced fuel cycle with partitioning followed by 'improved' conditioning of some selected radionuclides would substantially decrease the migration risk, but not the (potential) hazard. Separation of MAs before vitrification and special conditioning of these radionuclides into, for example, ceramic or crystalline matrices with a very low solubility in water is a possibility that would offer substantial advantages in the reduction of long term migration risk.

However, such practices do not reduce the radiotoxic inventory and its associated hazard. The reduction of the actinide inventory, or its mean half-life, is the only possibility to significantly reduce the long term hazard. To achieve this goal only nuclear processes are capable of modifying the nature of the isotopes involved and their associated half-lives or decay schemes.

### 2.1.6. Transmutation of minor actinides

The largest hazard reduction can be obtained through the fissioning (incineration) of the higher actinides, which decreases the intrinsic radiotoxicity by a factor of 100–1000 [1, 9]. This nuclear process is preferably carried out with fast neutrons, if available in excess, because of the neutron economy per fission. For incineration of MAs with thermal neutrons, the transmutation yield is higher, but the neutron economy is less efficient due to absorption. Even with P&T integrated in the nuclear fuel cycle, a substantial amount of radioactive material (activation products, MAs and traces of uranium and plutonium) will always accompany the bulk of the fission products.

Neutron irradiation of higher actinides also leads to neutron capture and the formation of other higher actinides. If the resulting half-life of these actinides is shorter, the radiotoxicity of the target increases proportionally. An example is the transmutation of $^{237}Np$, which leads to the generation of $^{238}Pu$, resulting in an increase of the radiotoxicity of the transmuted fractions that is proportional to the ratio of the half-lives ($2 \times 10^6$ and 87 years, respectively). In a waste management option the risk of this irradiation product, which is less soluble and shorter lived, is lower than that of the initial target, but the radiotoxicity is four orders of magnitude higher.

The reverse can also occur, for example with $^{244}Cm$ with a 18.1 year half-life, which is generated in high burnup fuel. After ten half-lives (~180 years), 99.9% of a separated curium target will have decayed into $^{240}Pu$, with a half-life of 6560 years. By transmutation of a $^{244}Cm$ target it would partially be transformed into $^{245}Cm$, with a half-life of 8500 years, having a slightly reduced radiotoxicity compared with the natural decay product $^{240}Pu$.

The licensing period for the proposed Yucca Mountain repository is 10 000 years. Partitioning of MAs followed by advanced conditioning is a complementary technology, which should be investigated. It has to be thoroughly examined to determine those radionuclides that benefit from transmutation and which transmuter would provide the cleanest end product. In the partitioning–conditioning scenario the partitioned fractions remain available for possible transmutation in the future, if this technology is fully implemented.

Transmutation with mainly fissioning is currently the most effective technology to reduce the actinide inventory and consequently its radiotoxicity and long term hazard.

### 2.1.7. Natural and archaeological analogues

Nuclear waste disposal is an extreme example of radionuclide ‘packaging’. One of the responsibilities of the nuclear industry is to demonstrate that an underground repository can contain nuclear waste for very long periods of time and that any releases that might take place in the future will pose no significant health or environmental risk. It must be taken into account that the engineered barriers that initially contain the waste will degrade and that some residual radionuclides may return to the surface in low concentrations at some time in the future due to groundwater movement through the natural barriers of the repository and due to environmental changes (e.g. climate changes and geologic phenomena).

One way of building confidence in engineered and natural barriers is by studying the processes that operate in natural and archaeological systems, and by making appropriate parallels with a repository. These processes are called natural analogues. Natural analogues are particularly relevant in the event that industrial quantities of conditioned waste are disposed of in deep underground structures. As an example, the geochemistry ruling their behaviour should be similar to that of uranium or thorium deposits in undisturbed geologic conditions.

Modern human-made barriers can provide confinement of all types of material for thousands of years, during which time the major fission products decay completely.

Further discussion is given in Annex I.

## 2.2. TECHNICAL ISSUES RELATED TO PARTITIONING AND TRANSMUTATION

Partitioning is to a certain extent a broadening to other radionuclides of the current reprocessing techniques that have been operating at the industrial level for several decades and for which the main facilities, at least in the European Union and Japan, exist or can easily be extrapolated from present day nuclear plants. Partitioning is a technology that can be considered to be a form of 'super reprocessing', in which uranium, plutonium and iodine ($^{129}I$) are removed during the processing; the MAs and the long lived fission products (LLFPs) ($^{99}Tc$ and $^{137,135}Cs$) would be extracted from the high level liquid waste (HLLW). Some LLFPs that are significant in long term waste disposal assessments (e.g. $^{93}Zr$, $^{107}Pd$ and $^{97}Se$) cannot be extracted unless isotopic separation is considered.

Partitioning itself does not create new radioactive substances; it deals with the same radionuclide inventory as traditional Purex processing. However, considerable complications may arise due to higher burnup or to shorter cooling times. Partitioning requires additional processing of spent nuclear fuel (i.e. new stages in the processing flowsheets and more complicated installations with resulting increased failure risks). Highly radiotoxic materials such as americium and curium obtained and separated in a concentrated form have high gamma and/or neutron emissions and are associated with complex shielding problems.

Partitioning generates more individual radionuclides (e.g. iodine, technetium and neptunium) or groups of radionuclides with analogous chemical properties, which can be managed in a safer way for long term intermediate storage or disposal (e.g. americium and curium).

A proper conditioning, resulting in a reduction of the solubility, is the most obvious approach. Incorporation into a stable matrix reduces the migration risk of accidental dispersion during the storage period and migration in the repository.

However, major changes in the safety situation will arise in a P&T fuel cycle, due to the transition to new fuels, which are generally much more radioactive both before and after irradiation, with higher levels of residual heat and helium production resulting from intense alpha activity.

Transmutation requires new fuel fabrication plants and irradiation technologies, which must be developed and implemented on an industrial scale. Existing nuclear power plants could in principle be used for transmutation, but many practical obstacles may arise (e.g. interference with the daily operation of the plants). New irradiation facilities such as dedicated FRs, accelerator driven transmutation devices and even fusion reactors have been proposed for transmutation–incineration purposes.

One of the main objectives of P&T has always been to reduce the long term hazard of spent fuel or HLW, this hazard being associated with the radioactive source term itself. In contrast, in the management of waste the long term radiological risk is stressed; this long term radiological risk is a combination of the potential hazard and the confining properties of the geologic media. The measures that have to be taken for hazard reduction are very different and much more fundamental than those for risk reduction.

For a desired reduction factor in total radiotoxicity of 1000, a target value of 99.9% has to be achieved for the recovery of each of the individual TRUs during the reprocessing–partitioning operations. With a 100-fold reduction in the TRU content the reduction in radiotoxicity could theoretically be reached after less than 1000 years.

It is obvious that a 100-fold reduction of the TRU mass in the waste compared with the OTC cannot be achieved in a single pass through a reactor. Multiple recycling will hence be necessary. In fact, the ideal P&T system should have a fuel cycle that is fully closed for the TRUs. Only the fission products would enter the waste stream, together with a 0.1% fraction of the cumulated and/or recycled TRU fraction. Given the limitations of irradiation facilities, such a system must be operated for many decades before equilibrium is reached in the core composition and in the radiotoxic output of the TRU losses.

## 2.3. EFFECTS OF CHANGES IN LONG TERM POLICY ON WASTE MANAGEMENT

The most important aspect associated with P&T is the long term implications of the short term decisions taken in the framework of nuclear waste management. Once a given path has been traced and the first steps have been made, it is difficult if not impossible to eliminate the consequences.

The OTC is reversible as long as an irreversible disposal in a deep geologic repository has not been realized (i.e. until the shaft is closed). As long as the repository is filled but not sealed, changing the decision taken is, while very expensive, not impossible within a period of, for example, 100–300 years. However, the fuel cycle facilities that are necessary to reopen the cycle would not be available and would have to be recreated. The case of the USA is to a certain extent a typical example of this situation. Civil reprocessing was abandoned in the early 1970s and a certain form of reprocessing (Urex and pyrochemistry) had to be reintroduced in order to eliminate TRUs from spent fuel.

The RFC as developed industrially in Europe reduces the plutonium content of HLW but transfers it to MOX fuel. The vitrified HLW contains the LLFPs and the MAs, which constitute the long term radiotoxicity but do not involve a criticality hazard. By transferring the plutonium into MOX fuel, some additional decades have been bought before the need arises to take decisions that have a long term impact. A significant reduction of the plutonium inventory can only be obtained if an important (~20–40%) fraction of the reactor fleet consists of FRs or equivalent fast neutron spectrum facilities. They do not currently exist and it will take several decades before a viable FR industry has been built up. The disposal of LWR MOX could therefore be delayed until the next generations decide how to equip their nuclear systems.

Since reprocessing of LWR $UO_2$ leads to the production of HLW, which contains MAs, prompt decisions have to be taken on how to proceed further. Separation of MAs has to be decided upon in the interval between reprocessing and vitrification; this period has been shortened in the large integrated reprocessing plants. Once the separation is performed the question arises of how to proceed with the separated radionuclides.

In most repository conditions $^{237}Np$ is the only MA with a very long term radiological impact, but $^{241}Am$, being the parent of $^{237}Np$, has to follow the same sequence of decisions. The radiological impact on the biosphere of $^{237}Np$ in vitrified HLW is non-existent over the physical lifetime of the glass (i.e. 10 000 to 100 000 years), but once conditioned as glass the radioactive waste cannot be recycled economically. Partitioning of MAs before vitrification and conditioning them into a stable ceramic form, followed by storage in retrievable conditions, is therefore a method to keep the decisions on this source term in the hands of future generations without increasing the risk. Transmutation of properly conditioned targets could in principle be undertaken in the future if dedicated burner reactors become available.

Reprocessing provides access to some fission products that are important radiological source terms. Existing technologies enable partitioning and conditioning of these radionuclides into appropriate matrices to be disposed of in suitable repository conditions if available.

The advanced fuel cycle with P&T incorporated is the most comprehensive approach. It requires dedicated fuel cycle and reactor facilities that go far beyond current nuclear technology. In particular, the transmutation approach calls for the development of FR burners and/or accelerator driven system (ADS) facilities, which may take 20–30 years to become industrially available. This option is the only one that offers a final solution to sustainable nuclear energy production.

A serious situation would occur if recycling were interrupted after, for example, 50 or 100 years. In this scenario the enrichment of plutonium in the nuclear energy generating plants would have reached an equilibrium level and the whole inventory must be disposed of at that time. Recycling of TRUs in a composite fleet of nuclear reactors, comprising LWRs and FRs, depends on a long term energy policy with a continuous political and economic backing of nuclear energy in the global energy mix.

The impact of this 'interruption' scenario on the design of a repository is far reaching: the radiotoxicity of the nuclear fuel streams after long term irradiation is multiplied by several orders of magnitude, the heat dissipation requirement is much higher and the effect of the long term radiotoxicity reduction is not attained for a period of several hundred years, as determined by the residual concentration and the decay time of $^{238}$Pu.

The most important decay chain [7] from a radiological point of view is the $4n + 1$ decay chain, comprising $^{245}$Cm, $^{241}$Pu, $^{241}$Am and $^{237}$Np. The $4n + 3$ decay chain includes, for example, $^{243}$Cm and $^{239}$Pu, and the $4n$ chain includes $^{244}$Cm and $^{240}$Pu (Figs 3 and 4).

In conclusion, the decision to operate the P&T fuel cycle should be supported over a sufficiently long period (70–100 years), until equilibrium is established between generation and consumption of TRUs, otherwise an interruption would imply multiplication of the radiotoxic inventory.

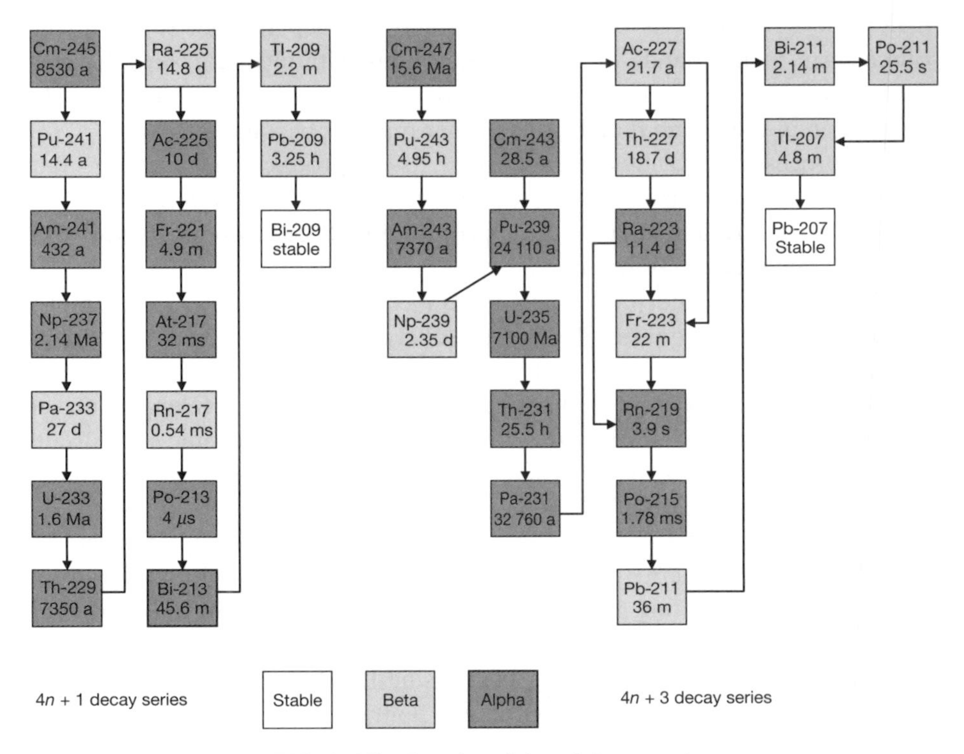


*FIG. 3. The 4n + 1 and 4n + 3 decay series.*

## 2.4. DECISION REQUIREMENTS FOR INTRODUCTION OF PARTITIONING AND TRANSMUTATION

### 2.4.1. Aqueous processing

As indicated in Section 2.2, partitioning can be considered as 'super-reprocessing'. Within the Purex process only uranium and plutonium are recovered at present, while neptunium, iodine and technetium could be separated by a modified Purex process. Other MAs, lanthnides, platinum group metals or gamma ray emitters such as strontium and caesium and other long lived radionuclides cannot be retrieved by the Purex process, and the use of specific extractants would need to be introduced.

If deep aqueous processing becomes technically achievable and industrially convenient:

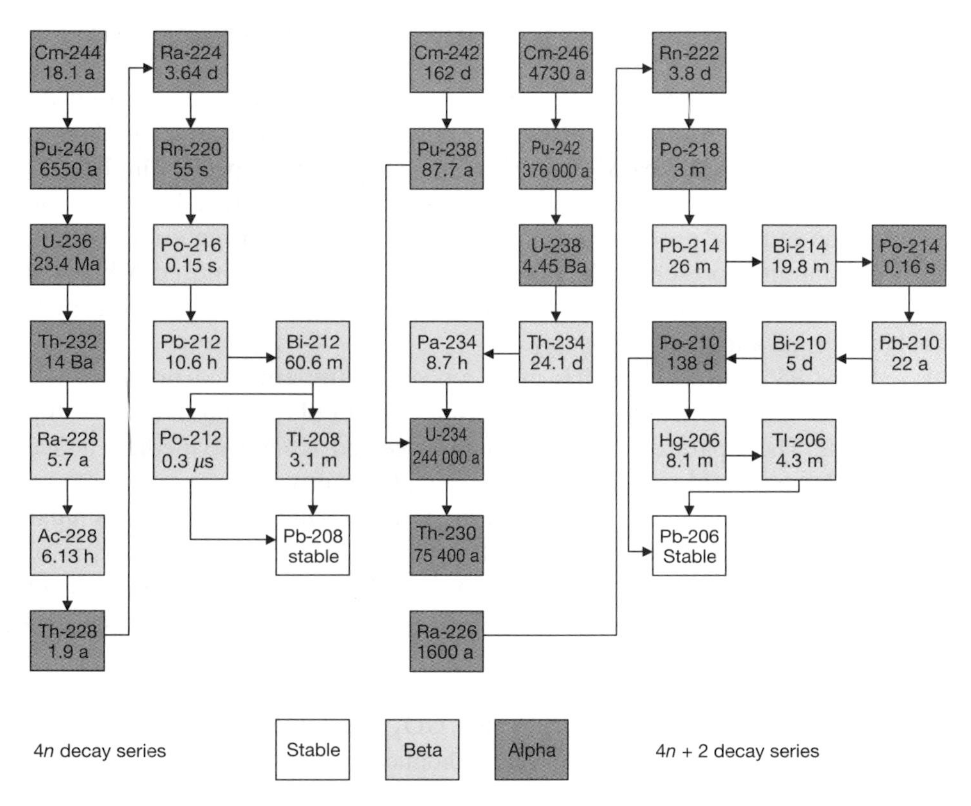


*FIG. 4. The 4n and 4n + 2 decay series.*

(a) An additional facility for the separation of MAs from high active raffinate (HAR) can be attached to an existing reprocessing plant.

(b) The separated MAs can be stored together with separated plutonium in existing storage facilities.

(c) The secondary waste resulting from the partitioning processes can be treated in the same facility in which low and intermediate level waste (LILW) is treated and conditioned.

(d) Iodine isolated from the dissolver off-gases can be conditioned and stored instead of discharged into the sea.

(e) New fuel fabrication plants with γ–n shielding will have to be erected when recycling in the fuel cycle is started.

### 2.4.2. Pyrochemical reprocessing and recycling of transuranic elements

The US Department of Energy has proposed a programme based on Urex and pyrochemical processes to optimize the cost and technological performance of the Yucca Mountain geologic repository. Special emphasis is placed on proliferation resistance measures. This philosophy is based on the elimination of those process steps in which sensitive nuclear material, particularly plutonium, is present in the pure elemental form. The proposed Advanced Fuel Cycle Initiative programme envisions partitioning and transmuting spent fuel in thermal and fast reactors [4].

Since 95% of spent fuel is made up of slightly enriched uranium, the elimination of uranium as fertile breeder material for plutonium is the first step to be made. The separated uranium will be processed as a low to medium active waste and transferred to a suitable burial site. Currently this process would be based on an aqueous technology called the Urex process, which is very similar to the conventional Purex process and would make use of the same type of extraction technology. After uranium removal, the remaining mixture of elements will be pyroprocessed for TRU burning in integral fast reactors (IFRs).

Japan has an activity [10] to establish an integrated fuel cycle of FRs with metallic fuels by pyrochemistry, in which $UO_2$ and MOX from LWRs can be treated by chemical as well as by electrochemical reduction. In that system all actinides are recycled in an FR fuel cycle, including TRUs produced in LWRs. Pyropartitioning of TRUs in HLW coming from a Purex facility is also envisaged. An ADS system with nitride fuel is proposed to transmute TRUs. Pyrochemistry would also be applied to recycle TRUs in this type of system. A collaboration agreement has been signed between the Japanese Central Research Institute of the Electric Power Industry (CRIEPI) and the European Commission Institute for Transuranium Elements (ITU) in order to launch a pyrochemical development programme using representative quantities of TRUs (see Section 3.5.4.1).

Extensive strategic evaluations [11] have been made in the Russian Federation to assess the technological impact of large scale transmutation on a sustainable nuclear programme.

### 2.4.3. Neptunium processing and transmutation issues

Chemical separation of neptunium in an LWR reprocessing plant using the Purex process is technically possible, but the implementation of this step in an industrial plant requires modification of the first extraction cycle flowsheet and connection to the HLW treatment complex. Once separated and purified, neptunium can be stored as an oxide; however, safeguards control is necessary, since it is in principle a fissile material.

Conditioning of separated neptunium for long term storage can be achieved by transforming it into, for example, a Synroc type of compound (a mixture of zirconolite, hollandite and perowskite), which is much more insoluble than glass forms. Long term surface storage or retrievable disposal in underground facilities are options to be studied.

Target fabrication for future transmutation is another alternative. For example, mixing of $NpO_2$ with aluminium powder and transformation into uranium/plutonium/neptunium subassemblies makes it suitable for irradiation in thermal and fast neutron flux reactors or in ADS facilities.

Irradiation by thermal neutrons [9] produces large inventories of $^{238}$Pu, while irradiation in FRs or in fast spectrum ADSs reduces this formation but does not eliminate it. Multiple recycling in dedicated FR or ADS facilities has to be considered to achieve complete incineration. Hot refabrication of $^{237}$Np–$^{238}$Pu targets is a technology that needs development in order to ensure a high decontamination factor (DF) from the ingrown actinides ($^{239}$Pu, $^{240}$Pu).

The following industrial facilities and processes will be necessary to produce and process the irradiation targets resulting from a reactor system of 100 GW(e) [1]:

(a) A neptunium chemical purification plant with a capacity of ~1.6 t/a;
(b) A neptunium target fabrication facility of an equivalent capacity;
(c) Interim safeguarded (or underground retrievable) storage of fresh neptunium targets;
(d) A dedicated chemical processing rig for irradiated targets;
(e) A hot refabrication plant for recycling of $^{237}$Np–$^{238}$Pu targets;
(f) An HLW connection with an LWR $UO_2$ vitrification plant;
(g) A medium level alpha waste treatment plant for effluent processing.

Long term storage or disposal of ~1.6 t of $^{238}$Pu and fission products (880 and 190 kW(th), respectively) will be necessary, depending on the fission rate and the cooling period.

A dedicated FR or ADS transmutation facility should be erected in the immediate vicinity. This complex ought to be associated with one of the big industrial reprocessing plants, in order to avoid unnecessary transfers and transport on public roads.

The real benefit of this development for waste management is the reduction of the neptunium inventory and of the very long term radiological risk associated with the migration of $^{237}$Np into the biosphere. However, it has to be remembered that quantitative transmutation of industrial quantities of separated neptunium inventories requires multiple recycling and hot chemistry processing of irradiated targets and ultimate storage over hundreds of years of significant quantities of $^{238}$Pu.

The radiotoxicity resulting from the transmutation increases dramatically and leads to an increase in decay heat release and in the use of specific conditioning methods to bridge the 500–800 year period necessary to let the $^{238}$Pu hazard decay to $^{234}$U.

### 2.4.4. Americium and curium processing and transmutation issues

The separation technology for americium and curium is still in the laboratory phase of development. Several techniques (see next section) have been tested in hot radioactive conditions, and many years of hot pilot scale tests will be necessary to realize a dependable industrial process.

The industrial implementation of americium and curium transmutation requires a highly sophisticated infrastructure. In the first stage, group separation of the americium–curium–rare earth fraction results in up to 50% rare earth, which has to be further purified in order to reduce the rare earth content to, for example, 10% or 5%, which is tolerable in an FR MOX reactor. Moreover, the americium and curium stream contains chemical impurities, which have to be eliminated in order to prepare a suitable target matrix for irradiation. Fully gamma and neutron shielded facilities (hot cells with heavy neutron shielding) are required to handle the effluent from the preliminary separation. The facilities to separate, condition and transmute the separated americium and curium stream are much more sophisticated than those for $^{237}$Np.

To implement this option, the following P&T plants and processes are necessary to treat annually the approximately 1.7 t of americium and curium discharged from a 100 GW(e) reactor system [1]:

(a) An $\alpha$–$\gamma$–n shielded americium and curium chemical purification plant:
  (i) An additional rare earth separation rig;
  (ii) Chemical purification of the americium and curium fraction;
  (iii) Optionally, americium and curium separation facilities.
(b) $\alpha$–$\gamma$–n shielded americium and curium target preparation.
(c) Interim storage of conditioned targets prior to irradiation.
(d) A transmutation facility for americium and curium based on ADS or FR technologies.
(e) $\alpha$–$\gamma$–n dedicated hot recycling rigs for irradiated targets.
(f) Disposal facilities for once through irradiated targets (~620 kW(th)).
(g) Optionally, a dedicated curium storage facility for $^{240}$Pu recovery.

(h) Connection to HLW vitrification.
(i) A high and medium level alpha waste treatment plant.

Americium and curium targets contain significant amounts of $^{238}$Pu, $^{244}$Cm and fission products; because of this combination, the irradiated capsules are very difficult to handle and reprocess. One alternative to spent target reprocessing consists of carrying out a once through irradiation in a thermal island of an FR MOX reactor until 200 dpa (limit of capsule/target pin cladding) in order to eliminate further processing [9].

According to this scenario, 98% of a given americium inventory could be eliminated by neutron irradiation over ~20 years. The irradiated capsules from such a test, if they resist the intense radiation and the high helium pressure, may have to be overpacked before their long term storage to avoid leakage inside the storage facilities. The transformation of the americium target into a mixture of 80% fission products and 20% TRUs is possible in a target type capsule but cannot be extrapolated to industrial quantities, since in this case the global impact of the americium depletion in the target and the americium generation in the driver fuel has to be taken into account.

Transmutation of americium and curium is a technical challenge in its operational phase, but the radiological impact is slightly positive (i.e. an actinide reduction factor of 10–20 can be expected) and its influence on waste management is rather limited. Once an inventory has been transformed (almost) completely in a fission product mixture, the management of it is identical to the management of actinide free vitrified HLW. The main long term benefit is the elimination of $^{241}$Am, which is the parent of the very long lived and slightly mobile $^{237}$Np, but its partial (~12%) transformation into some long lived plutonium ($^{238,239,240}$Pu) and curium isotopes is a drawback that limits its usefulness for long term waste management.

It is obvious that ADS systems could be more efficient at providing the necessary neutrons to transmute a fertile $^{241}$Am–$^{243}$Am mixture than critical FRs, in which the driver fuel will itself be a generator of americium and curium. From the safety point of view, the loading of the ADS reactor is more flexible, since the void reactivity coefficient remains negative even with very high TRU loadings. It is beyond the scope of this report to discuss the reactor safety implications of this new type of transmutation facility [2].

### 2.4.5. Transuranic element processing and transmutation issues

The presence of plutonium together with MAs determines the throughput and criticality requirements of the processing facility. The mass of the separated TRUs is roughly 10–15 times higher than the mass of the separated MAs. The criticality limitations are determined by the plutonium content and by the nature of the extraction process. The use of aqueous extractants greatly limits the throughput and specific capacity of a processing facility. By using pyrochemical methods and excluding aqueous solutions during the treatment of separated TRUs, the specific volume of the treatment plant can be greatly reduced and high burnup fuel with a high decay heat emission can be processed.

The isolation of TRUs from spent nuclear fuel makes use of a combined aqueous–pyrochemical separation process. The bulk of uranium is extracted by tri-n-butylphosphate (TBP) separated from the TRUs and discharged as medium level waste or stored for future use. The HLLW is calcined and dissolved in a molten salt bath. The adjacent pyrochemical plant separates the TRUs from the bulk of the fission products. The TRUs are purified by electrorefining and reductive extraction and transformed into metal or nitride. The TRU metal mix is recycled by high temperature casting in fuel pins. The recycled fuel pins are irradiated in FRs or dedicated accelerator driven subcritical reactors. The following facilities and processes are required:

(a) A mechanical head end and TBP extraction facility (Urex);
(b) Calcination of HLLW;
(c) Metal oxide reduction;
(d) Electrorefining or a reductive extraction process facility for TRUs in molten LiCl–KCl salt at 700°C;
(e) Fission product elimination and conditioning as zeolite embedded in glass;
(f) A metal fuel or nitride fuel fabrication facility for very hot fuel handling ($\alpha$, n, $\gamma$);
(g) A dedicated ADS facility for incineration by multiple recycling of TRUs.

None of these facilities have been constructed at the industrial level, except those for the calcination step, and most of the listed technologies are still in the design or laboratory phase of development. However, a electrorefiner has been tested with real fuel mixtures at Argonne National Laboratory–West.

# 3. NON-PROLIFERATION ASPECTS OF PARTITIONING AND TRANSMUTATION

As previously discussed, P&T has been considered by some States as a technical option for long term HLLW management strategies. P&T technologies involve the separation (partitioning) of MAs such as neptunium and americium from HLW to reduce its radiotoxicity. Thus the application of P&T techniques to a nuclear fuel cycle could result in increased inventories of separated neptunium and americium, which, prima facie, could pose a proliferation risk, because these materials could be used for nuclear explosive devices. At the very least, partitioning and subsequent transmutation of these materials through irradiation in dedicated nuclear facilities (FRs or ADSs) would certainly have an impact on safeguards implementation at facilities involved in P&T. Given the potential proliferation risk associated with a P&T strategy, the non-proliferation aspects of P&T should be carefully addressed at an early stage of development.

An evaluation of the proliferation risks potentially posed by P&T developments and applications and of their impact on IAEA safeguards implementation, and studies of proliferation resistance measures, needs to be an integral part of the investigation and assessment of P&T strategies and techniques.

## 3.1. PROLIFERATION POTENTIAL OF NEPTUNIUM AND AMERICIUM

It has been recognized for many years that some TRUs other than plutonium, in particular neptunium and americium, if available in sufficient quantities, could be used for nuclear explosive devices. Their respective critical masses are estimated to be of the same order as those of some other fissile actinides (i.e. ~50 kg). Owing to its long half-life (2 140 000 years), neptunium has no heat or radiation emissions that would complicate its use in a nuclear explosive device. Americium, however, produces high levels of heat and radiation, which would greatly complicate its use in a nuclear weapon; it would therefore require considerably more skill and resources to handle and use americium to manufacture a nuclear device. Other transuranics formed in fuel during the operation of a nuclear reactor (e.g. curium, berkelium and californium) also have fissionable isotopes. However, their more limited availability, high thermal output, short half-lives and other nuclear properties make them unsuitable for use in nuclear explosive devices. The results of an extensive survey of the isotopes of TRUs carried out by nuclear weapon States led to the conclusion that no elements other than neptunium and americium are likely to pose a proliferation potential for at least several decades.

Neptunium and americium, however, are not covered by the definition of special fissionable material in the IAEA Statute, since the availability of meaningful quantities of separated neptunium and americium was considered remote, and their detailed consideration was not warranted for safeguards purposes, at the time the Statute was adopted in 1956. In 1971 the IAEA Director General's Standing Advisory Group on Safeguards Implementation, in its advice to the Secretariat regarding 'threshold amounts', noted that, in addition to plutonium and high enriched uranium, "the assembly of a nuclear explosive using neptunium or americium may also be physically practical; however, given the requirements to produce these materials and the quantities existing in commerce, their detailed consideration is not warranted for safeguards purposes at this time."

There have since been technical developments that include the application of P&T technologies to long term waste management. In addition, neptunium and americium have commercial uses, which could provide incentives for their separation and recovery. Although not suitable for use as fuel in thermal reactors, neptunium and americium can be used as fuel in FRs. Waste management programmes involving P&T, in addition to the commercial incentives for isotope recovery, would clearly increase the availability of separated neptunium and americium worldwide.

## 3.2. MONITORING SCHEMES FOR NEPTUNIUM AND AMERICIUM

Against this background information and given also the inherent proliferation potential of neptunium and americium, certain controls should be applied when it becomes necessary to deal with their proliferation risks.

Until 1999, separated neptunium and americium were not subject to any international controls or monitoring schemes that could foster confidence that the relevant materials were not being used in the manufacture of nuclear explosive devices. Some degree of intergovernmental control was provided by a 1994 memorandum of understanding signed by France, the Russian Federation, the UK and the USA, in which these nuclear weapon States agreed to establish export controls on neptunium. These controls were expanded in 1998 with two common understandings involving France, Japan, the UK and the USA to cover controls on neptunium and americium. In addition, the Wassenaar Arrangement, co-founded by 33 States in 1996 to contribute to regional and international security and stability, provides that transfers of separated neptunium in quantities greater than 1 g should be controlled.

As a result of the increasing awareness of the proliferation potential of neptunium and americium, and of emerging projects in peaceful nuclear programmes that could lead to an increase in the available quantities of separated neptunium and americium, the IAEA Director General provided a report to the Board of Governors on "The Proliferation Potential of Neptunium and Americium" in November 1998 [12].

During the Board's discussion, a spectrum of views on the proliferation risks of neptunium and americium was reflected by Member States. In September 1999, after extensive deliberations, the Board agreed, as included in the Chairman's conclusion, that "the proliferation risk with regard to neptunium is considerably lower than that with regard to uranium or plutonium"; and the Board believed that "at present there is practically no proliferation risk with regard to americium." The Board endorsed the implementation of monitoring schemes for neptunium and americium through which the Secretariat could provide assurance that the quantities of separated neptunium and americium in States with comprehensive safeguards agreements (CSAs) remain insufficient to pose a proliferation risk and which provide timely notification to the Board if this situation were to change [13].

Given the difference in the proliferation risks posed by neptunium and americium, flowsheet verification (FSV) was introduced as an element of the monitoring scheme to cover the acquisition path of separated neptunium from indigenous production. FSV consists of the following: a set of confirmative measures such as examination and verification of design information relevant to neptunium; examination of relevant process records and monitoring of key process parameters; measurement of randomly selected samples; application of containment and surveillance measures; and use of environmental sampling and analysis. These measures, if implemented at candidate facilities in CSA States, would provide direct confirmation that these facilities are being operated as declared with respect to the recovery or separation of neptunium. It should be noted that FSV is basically a qualitative approach and does not entail detailed material accountancy of neptunium.

Candidate facilities for the application of FSV include those that have an actual or potential capability to separate appreciable amounts of neptunium, including in reprocessing plants, MOX fuel fabrication facilities with plutonium conversion operations and/or wet scrap recovery, conversion facilities involving neptunium and HLLW vitrification facilities. Candidate facilities also include large scale laboratory facilities engaged in R&D associated with developing actinide partitioning technologies. The monitoring of such facilities would be important because the equipment used and the experience gained in their operation would be directly relevant to the separation of larger quantities of neptunium.

Specific FSV activities to be carried out at any given facility would depend on the type of facility, the scope of operations in question, the amounts of relevant material associated with that facility and the layout, equipment and operating procedures used at the facility. There would be considerable flexibility in the selection of appropriate measures and in the level of intensity with which any individual measure would be applied, subject to consultations with the facility operators and State authorities. However, it can be foreseen that a reprocessing plant would require the most extensive application of all of the FSV activities, because a reprocessing facility presents the most extensive and complex possibilities for separating neptunium from the main process streams. A large scale R&D facility would also present the same complex possibilities for neptunium separation as a reprocessing plant, but on a smaller scale. All possible FSV measures may have to be carried out at such a facility.

These monitoring schemes have started to be implemented with the voluntary cooperation of relevant States in order to monitor international transfers and stocks of separated neptunium and americium in CSA States. Technical parameters and modalities for monitoring the production of and trade in separated neptunium and americium have been defined for their effective application.

## 3.3. IMPACT OF PARTITIONING AND TRANSMUTATION ON NUCLEAR NON-PROLIFERATION

Development of P&T could lead to an increase in the available quantities of separated neptunium and americium. Industrial applications of a P&T cycle in the future might involve substantial amounts of these materials. The impact of P&T on nuclear non-proliferation therefore needs to be carefully addressed in a feasibility study of P&T strategic and technical options. In practice, the proliferation risk and associated verification scheme would presumably depend on not only the quantity but also the quality of these materials, which would be determined by the partitioning schemes being deployed in a P&T cycle. On the basis of a study performed by the IAEA Secretariat during the development of proposals for the IAEA Board of Governors to consider in order to respond to the proliferation potential of neptunium and americium, a preliminary assessment can be made of a P&T cycle that would involve separated neptunium and/or americium.

### 3.3.1. Short and medium term impact

In the short term, the currently designed monitoring scheme for neptunium and americium would be a cost effective means of providing assurance that the quantities of separated neptunium and americium in CSA States, including those resulting from P&T development programmes, are, and remain, small. Under this regime, large scale laboratory facilities engaged in developing actinide partitioning technology would be subject to FSV in order to maintain confidence. It would also enable the IAEA to inform the Board of Governors in a timely manner whenever the accumulation of separated neptunium and americium in a CSA State was about to become substantial. However, strengthened monitoring arrangements at specific facilities or in a specific CSA State might become necessary in the medium term, when technological advances in partitioning have been made and separation activities have become significant. Strengthened arrangements could include an increase in the intensity of and an extension of the FSV activities as currently designed, and possibly an expansion of FSV to include americium, in order to deal with the changed situation, and would require close cooperation of the State and facility operators involved.

### 3.3.2. Long term impact

The long term impact of P&T on nuclear non-proliferation could become significant and extensive when the application of a P&T cycle reaches the industrial scale. It would involve institutional arrangements (political, legal and commercial) and IAEA safeguards.

#### *3.3.2.1. Impact on institutional arrangements*

If the development of P&T resulted in industrial scale applications, and substantial quantities of separated neptunium and americium became available, it would certainly have a significant impact on nuclear non-proliferation institutions, since the monitoring schemes as currently designed or subsequently strengthened would be insufficient to deal with such circumstances. It seems likely that the application of IAEA safeguards with detailed material accountancy would be necessary for such industrial scale applications of P&T. In order for neptunium and americium to become subject to safeguards, these materials would have to be included in the statutory definition of special fissionable material, which could be done by a determination to that effect by the IAEA Board of Governors. Each party to a CSA would need to accept the change in the definition of nuclear material in its safeguards agreement. Consequential amendments to other provisions of safeguards agreements would also be necessary, as would modifications to existing subsidiary arrangements. It would be a lengthy negotiation process for both State parties and the IAEA.

As a result of the amendment of the definition of nuclear material, a State party to the safeguards agreement would be legally bound to accept safeguards on neptunium and americium. Thus the requirements of the safeguards agreement relating to record keeping, reporting, inspection and international transfers would apply to neptunium and americium, as would the provisions of the agreement relating to non-compliance. Consequently, the IAEA would be able to provide the same degree of assurance regarding the use of neptunium and americium as it is able to provide under safeguards regarding the use of the materials currently included in the statutory definition of special fissionable material.

#### *3.3.2.2. Impact on safeguards implementation*

The amendment of the definition of nuclear material would entail additional activities on the part of States and the IAEA. States' accountancy reports for nuclear material inventories at reactors and other spent fuel storage locations would have to include the quantities of neptunium and americium contained in fuel. States' accountancy reports for reprocessing, waste treatment and plutonium facilities would have to include these materials, which would require increases in the sampling and laboratory analyses performed by operators in order to prepare these reports. While the small quantities of separated neptunium and americium known to exist could be exempted from safeguards in accordance with existing provisions in safeguards agreements, this would entail additional paperwork for both sides. The IAEA would also need to conduct the necessary report processing, inspection and sample analysis activities. In addition, the IAEA's systems for safeguards information treatment, nuclear material accounting and record keeping, inspection support, and evaluation and reporting would require modification, all of which would require substantial resources and staff effort.

#### *3.3.2.3. Safeguards implementation at future partitioning and transmutation cycle facilities*

As discussed in the previous sections, future P&T cycles might consist of:

(a) Partitioning facilities applying an aqueous or pyrochemical process;

(b) MA conditioning or TRU fuel (target) fabrication facilities, which would be new to the nuclear industry;
(c) Irradiation facilities such as conventional reactors (the least likely option), dedicated burner reactors or ADSs, which are still in conceptual development and might take several decades before they are industrially available.

It has been foreseen that most, if not all, of these facility types would be new in terms of safeguards implementation. Application of safeguards to these facilities would entail substantial efforts in the development of safeguards approaches and new methods and techniques to detect and deter diversion of relevant material, misuse of facilities or undeclared relevant material and activities. Developing new analytical techniques or improving existing analytical techniques as required for material accounting of neptunium and americium would be major challenges in these efforts. Although an advanced aqueous process has been demonstrated at the laboratory scale for partitioning, most of the other P&T technologies are still at the stage of conceptual development, with many uncertainties and open issues. It is too early at present to make detailed considerations of the safeguardability of the envisaged P&T cycle. More important issues are how much it would cost and how much effort it would entail, not only regarding the costs directly related to safeguarding P&T in the States involved, but also the additional costs to operators and to IAEA safeguards implementation as a whole.

## 3.4. DEVELOPMENT OF PROLIFERATION RESISTANT PARTITIONING AND TRANSMUTATION TECHNOLOGY

Having recognized the potential proliferation risks associated with the development and application of P&T techniques, and the potentially significant impact of P&T applications on IAEA safeguards, proliferation resistance of P&T needs to be considered at an early stage of development, and the development of proliferation resistant P&T technologies should be encouraged.

### 3.4.1. Proliferation resistance measures

Proliferation resistance could result from a combination of the institutional (political, legal and commercial) arrangements governing the implementation of P&T, the technological features incorporated in a P&T cycle and the verification provisions to be applied to it.

Principles and guidelines for proliferation resistant nuclear energy systems are being developed by the IAEA Secretariat under the International Project on Innovative Nuclear Reactors and Fuel Cycles (INPRO), some of which could be applied to P&T systems. The following principles may be considered as examples:

(a) Optimizing P&T development and application strategies with the aim of reducing proliferation risks;
(b) Reducing the availability or accessibility of weapons usable material;
(c) Avoiding substantial stockpiling of separated neptunium and americium;
(d) Incorporating proliferation resistance features in the designs and operational modalities of P&T facilities;
(e) Enhancing effectiveness and decreasing costs, with the aim of facilitating IAEA safeguards implementation.

### 3.4.2. Proliferation resistant strategies for partitioning and transmutation development

To date, partitioning of MAs has been demonstrated on the laboratory scale. TRU fuel or target fabrication and transmutation still remains in conceptual development. As outlined in the previous sections, industrial applications of these techniques would likely be at significantly different times, and partitioning might be implemented on an industrial scale well before transmutation. This scenario could result in stockpiling of substantial quantities of separated MAs, which will have to be stored while waiting for the availability of transmutation capabilities. Such a scenario would benefit early disposal of HLW, but would significantly increase, in addition to the safety and security concerns, proliferation risks and verification costs. Synchronization of industrial scale applications of partitioning with transmutation would therefore apparently be a proliferation resistant strategy for P&T developments and applications.

It seems unlikely that each State with an interest in P&T would have its own P&T cycle. For that reason, the implementation of P&T through international P&T fleets could be an attractive strategy for the purpose of proliferation resistance.

### 3.4.3. Intrinsic technological measures

Inherently proliferation resistant technologies that could possibly be incorporated in process and facility designs are clearly preferable to extrinsic measures (institutional arrangements and safeguards), and should be investigated, developed and incorporated in P&T.

A few suggestions have been made regarding proliferation resistant features that could reduce the availability or accessibility of weapons usable material by controlling the quality of the material concerned (e.g. isotopic composition, radiological properties and chemical forms). For example, a partitioning process could be so designed that it would only separate MAs from HLW in a mixture with selected LLFPs. It could also be so designed that MAs are not mutually separated. Products from such processes would not be directly suitable for weapons purposes and would be self-protected by the high levels of heat and radiation emissions. A strengthened FSV variant might be applied as a confirmatory measure at such partitioning facilities, while safeguards implementation would be needed only if and when pure materials were produced.

Proliferation resistant features could also be incorporated in the designs and operational modalities of the facilities to be used for P&T. These might include built-in engineered barriers to prevent the diversion of the materials concerned or an attempt to misuse the facility, providing a means to detect the diversion or misuse at an early stage, or making modifications to the flowsheet extremely difficult or impossible. Such facility designs might also stipulate strict operational conditions and parameters; any deviation from such normal conditions would cause serious safety or radiation risks.

## 3.5. TECHNICAL ASPECTS OF PROLIFERATION CONTROL

The proliferation risks associated with the fuel cycle supporting the nuclear power industry have been investigated and adequate safeguards measures have been fully implemented. In the European Union the safeguards authorities of Euratom and the IAEA carry out inspections and perform verification measurements on key material. Safeguards measures are sufficiently mature to provide assurance that nuclear material cannot be diverted without being detected. In contrast, however, dedicated P&T technology is still at an early stage of development. Consequently, proliferation resistance and safeguardability are issues that need to be evaluated carefully.

Some aspects of MA P&T will be discussed in connection with proliferation resistance and safeguardability assuming that the P&T activities will be separated from the main nuclear power industry and carried out in dedicated facilities, the aim of which is the reduction of the long lived waste inventory. This option, called the ‘double strata’ approach, will be used as an example in the discussion of the safeguards issues.

### 3.5.1. Partitioning

The aim of partitioning is the separation of the MAs (and possibly selected LLFPs) from spent nuclear fuel. This separation can be achieved either by an aqueous process or by a dry process.

Hydrometallurgical and pyrochemical reprocessing should not be considered as competing but rather as complementary technologies. In the double strata concept [14] the Purex process (first stratum) would be extended with an additional separation of radiotoxic elements from the raffinate or HLLW. This should be achieved by advanced aqueous partitioning. In the following transmutation cycle (second stratum), pyroreprocessing should be used:

(a) Aqueous (hydrometallurgical) partitioning can start from the HAR resulting from the first extraction cycle of Purex reprocessing. The presence of the fission products is a guarantee of proliferation resistance, and the plutonium content is very low. Industrial implementation is necessarily connected to large reprocessing plants. MA separation by aqueous methods reduces considerably the mass of the separated fractions that could be treated in dedicated facilities elsewhere.

(b) Pyrochemical partitioning proceeds from a HAR containing the TRUs that need to be transformed into an anhydrous (salt, metal, nitride) form. The bulk of MAs and fission products is treated by electrorefining processes by which a gross separation between the bulk of the fission products and the MA containing fraction is realized. Subsequent transmutation in dedicated irradiation facilities (second stratum with FRs or ADSs) requires specially designed fuel cycle facilities and the use of pyrochemical processes.

### 3.5.2. Proliferation resistance of advanced aqueous processing

As pointed out by Bragin et al. [15], the proliferation resistance of a fuel cycle can be increased by reducing the strategic value of the material and by design features that prevent diversion. In the separation of MAs from HLLW, only material of low plutonium content is processed. The processes are not designed to fully separate lanthanides from MAs. Among the lanthanides a number of radionuclides are found with high neutron capture cross-sections. Furthermore, some MAs are not mutually separated (e.g. americium and curium remain in the same fraction). The presence of the lanthanides and americium leads to a high dose rate arising from the material.

The material is, owing to its isotopic composition and high radiation level, not directly suitable for weapons purposes and requires further chemical processing. This would involve longer warning times and requires a high technological standard in order to prepare a material suitable for weapons purposes. Consequently, a certain degree of proliferation resistance is inherent in this scenario.

### 3.5.3. Safeguardability of advanced aqueous partitioning

HAR typically contains 0.09 g of americium and curium per litre, 0.10 g of neptunium per litre and less than 0.006 g of plutonium per litre. At present, the available measurement technology allows for plutonium assay at this level of concentration after appropriate sample preparation. The MAs, however, also need to be controlled. A typical reprocessing plant with an annual throughput of 800 t of spent fuel is estimated to produce some 4000 $m^3$ of HAR per year. An advanced aqueous partitioning of the required capacity of 4000 $m^3/a$ would then show an annual throughput of about 25 kg of plutonium, 410 kg of neptunium and 370 kg of americium and curium. A measurement uncertainty of 5% relative should hence enable detection of the diversion of a quantity equivalent to the respective critical masses of neptunium or americium (~50 kg). Measurement techniques for neptunium and americium have been developed [16], and more recently advanced measurement techniques for directly determining neptunium in highly active solutions have been successfully tested [17]. These methods can easily be applied to advanced aqueous reprocessing and should allow a (near) quantitative verification of material flows and inventories. Evidently, these concentration measurements need to be complemented by tank volume measurements. Advanced aqueous reprocessing facilities will be comparable with existing chemical processing facilities. Consequently, containment and surveillance techniques can be applied in analogy, complementing the aforementioned measures.

The separation of MAs from HLLW is a much more difficult task, since the concentration of nitric acid is close to 2M and the final salt content is close to crystallization. Although much R&D work has in the past focused on this type of liquor, it is not suggested to pursue this route. Once concentrated from 5 $m^3/t$ to ~0.3 $m^3/t$ (concentration factor of 15) the liquor is ready for vitrification and is not suitable for delicate chemical operations. A DIAMEX (diamide extraction) flowsheet for the partitioning of MAs from high active concentrate is under development within the European PARTNEW programme (2000–2003). A successful hot test of DIAMEX high active concentrate was carried out in the summer of 2003 at the ITU in Karlsruhe, Germany.

### 3.5.4. Partitioning by pyroprocessing

Pyrometallurgical processing (pyroprocessing) to separate nuclides from a radioactive waste stream involves several techniques: volatilization, liquid–liquid extraction using immiscible metal–metal phases or metal–salt phases, electrorefining in molten salt, fractional crystallization, etc. They are generally based on the use of either fused (low melting point) salts such as chlorides or fluorides (e.g. LiCl + KCl or LiF + $CaF_2$) or fused metals such as cadmium, bismuth or aluminium.

Pyroprocessing can readily be applied to high burnup fuel and fuel that has had little cooling time, since the operating temperatures are high. However, such processes are at an early stage of development compared with hydrometallurgical processes, which are already operational.

Separating (partitioning) the actinides contained in a fused salt bath involves electrodeposition on a cathode, extraction between the salt bath and a molten metal (e.g. lithium) or oxide precipitation from the salt bath.

(a) Pyroprocessing installations can be highly compact and directly connected to the reactor operation.
(b) Partitioning, fabrication and irradiation could be carried out in an integral unit and thus the present difficulties encountered with frequent nuclear material transport could be eliminated. As a consequence, a considerable cost reduction could possibly be expected.
(c) Owing to the higher radiation resistance of the proposed processes in molten salts, the reprocessing of short cooled spent fuel is possible. Cooling times of the spent fuel as short as a few months seem possible, compared with the present three to seven years and longer needed for aqueous reprocessing.
(d) The process is more proliferation resistant than aqueous reprocessing because the electrochemical potentials of actinides and lanthanides are very close, and fissile material cannot be separated in pure form. Plutonium, for example, is co-deposited together with MAs and some (highly active) lanthanides.
(e) Criticality problems are less severe, since in dry processes neutrons are less strongly thermalized. The consequence of this is that higher concentrations of actinides can be handled.

One of the key features of pyroprocessing is that it results in impure plutonium, which is not well qualified for making nuclear weapons. The plutonium removed from the salt contains some uranium, other TRUs and some fission product contamination. It is so contaminated and highly radioactive that it would not be suitable for the construction of a nuclear weapon. Similar observations have already been made for the IFR fuel cycle [18], in which the material is considered to be self-protecting for the reasons mentioned above. The arguments put forward in Section 3.4.3 on the lack of 'attractiveness' of the product (isotopic quality, chemical purity, heat output and radiation) also apply.

### *3.5.4.1. Safeguardability of pyroprocessing*

According to the present safeguards inspection system, effective safeguarding of pyroprocessing plants would be more difficult than is the case for conventional plants, since it is more difficult to measure and keep track of the fissile material in the process. The ITU and CRIEPI have been working on methods for the quantitative analysis of plutonium and MAs in salt and cadmium matrices.

In order to establish a material inventory, a bulk measurement (volume or mass) must be taken, followed by a sample from the bulk and a measurement of the compound concentration in the sample. Whereas the fissile material inventory can be determined by bulk measurement of the initial metal alloy mass of spent fuel, sampling of the solidified electrolyte or of the metal cathode is much more difficult than in aqueous processes. At this early stage of process development it has been observed that the material sometimes suffers from lack of homogeneity, hence a higher number of samples have to be taken for analysis. Effective safeguards verification of fissile material at new types of pyroprocessing plant has to be developed.

With respect to analytical methods, well established chemical methods such as isotope dilution mass spectrometry (IDMS) or inductively coupled plasma mass spectrometry (ICP–MS) can be applied for the assay of plutonium and MAs. Chemical sample preparation will be required, whichever of the two analytical methods is applied. It is, however, relatively labour intensive and requires a careful study of the chemical recoveries of the critical elements.

IDMS cannot be applied to neptunium, due to the lack of an appropriate spike isotope. Americium and curium can be measured by IDMS using $^{243}$Am and $^{248}$Cm as spikes. ICP–MS can be applied to all MA elements. However, separation chemistry (e.g. ion exchange, extraction chromatography or high performance liquid chromatography) is still required to avoid isobaric interferences.

Better radiometric methods can be used for the assay of MAs in samples from pyroprocessing. The efforts required for sample preparation can be reduced to a minimum. Development work in this context is being carried out at the ITU in order to systematically study and establish an analytical methodology.

Quantitative assay of uranium, plutonium and MAs using non-destructive methods (i.e. samples from the process are measured without further treatment) may suffer from strong matrix effects and from the presence of an overwhelming mass of fission product isotopes. Nevertheless, some non-destructive techniques might be applicable for measuring the spent fuel and for monitoring individual process parameters, but these will need to be further developed. These measures need to be complemented by the traditional techniques of containment and surveillance, by design information verification and by monitoring of essential system parameters.

In case of loss of continuity of knowledge and in order to verify that essential system parameters have not been altered with the intention to obtain pure products (plutonium or neptunium), two analytical approaches are possible. First, swipe samples taken from inside a hot cell could be checked by particle analysis methods for the presence of high purity material. Second, the presence of pure plutonium or pure neptunium on the electrode would indicate an anomaly. In either case only elemental analysis needs to be performed.

## 3.6. SAFEGUARDS AND PROLIFERATION ISSUES OF TRANSMUTATION

### 3.6.1. Proliferation resistance of transmutation reactors

Proliferation resistance of transmutation reactors with their novel technical features needs to be addressed at an early stage of development.

Proliferation risks associated with ADS technology have been reviewed at the ITU and critically discussed in a series of articles [19–22], with emphasis on the potential proliferation aspects of both high power accelerators and spallation sources. It has been known since the 1940s that bombardment of a uranium target with high energy protons or deuterons would produce a large yield of neutrons and that these neutrons can be used in turn to produce fissionable material through nuclear reactions. G.T. Seaborg produced the first human-made plutonium using an accelerator in 1941.

It has been shown that relatively small commercial cyclotrons (150 MeV, 2 mA) are capable of producing amounts of fissile material greater than the 'screening limit'. The IAEA screening limit is a capability to produce 100 g of plutonium per year or to operate continuously at thermal powers greater than 3 MW. However, the accelerators foreseen for accelerator driven waste transmutation can be expected to produce tens of kilograms of plutonium per year.

It has been shown that this technology is now mature enough for fissile material production and that there is a need to issue regulations on this subject. The US Department of Energy announced regulation changes on this subject in a federal announcement issued on 27 March 2000 (10 CFR Part 810, RIN 1992-AA24). In summary, these regulations restrict the export of accelerator driven subcritical systems and their components capable of continuous operation above 5 MW(th). Although accelerators are not specifically mentioned, there is some room for interpretation as to whether an accelerator is a component of a 'utilization facility'.

### 3.6.2. Safeguardability of transmutation reactors

Accelerator driven power units or waste burners will be subject to international safeguards provided that any of the following conditions are satisfied:

(a) The fuel or target material consists of thorium, uranium or plutonium, which are already subject to IAEA or Euratom safeguards;
(b) Normal operation, misuse or clandestine operation of the accelerator could be used for the production of significant quantities of fissionable material.

As far as fresh or irradiated fuel is concerned, new verification techniques may need to be developed in order to permit verification of receipts to the ADS and shipments of nuclear material from the ADS. If the nuclear material becomes difficult to access for verification during the annual physical inventory verification, containment and surveillance methods will need to be applied to maintain continuity of knowledge of the inventory. Such systems exist and provide effective and efficient safeguards verification in different reactor types.

It is important that safeguardability criteria be included in the design of the ADS from the beginning, in order to ensure cost effectiveness.

The production of fissionable material, either during normal operation or by clandestine operation, is a feature that is already available in existing reactors. A detailed diversion analysis needs to be performed for an ADS based on the design parameters and layout of the system. Following this analysis, measures will be identified and implemented that would ensure a high rate of detection of misuse, probably within the relevant detection time.

The most important safeguards measures for an ADS would be containment and optical surveillance devices and monitoring systems. These systems allow freezing parts of the ADS such that any access for misuse or reconfiguration would be detected. Highly tamper resistant monitoring devices permit real time observations of important operational or system control data and would therefore indicate any deviation from normal operational practices.

The fact that the neutron source in the ADS is outside the reactor is a new feature but does not present a particular problem; it requires that the neutron source be included in the safeguards system such that any attempt to misuse or redirect the beam to a non-controlled location of targets would be detected. During the shutdown period of an ADS, assurance of 'non-breeding' could, for example, be obtained by a tamper resistant control of the power of the accelerator.

Depending on the detailed layout and operational characteristics of such a facility, and the relationship with the fuel fabricator or spent fuel receiver, the safeguards effort could be comparable with that of certain types of critical reactor.

# 4. FUEL CYCLES

## 4.1. INTRODUCTION

### 4.1.1. Once through fuel cycle

The OTC scenario (Fig. 5) is the basic spent fuel management option in Canada, Spain, Sweden, the USA and some other States. Given the present low uranium price, the OTC provides the lowest cost nuclear energy production. However, it implies that the residual fissile material content (1% Pu and 0.8% $^{235}U$), as well as the remaining fertile material content ($^{238}U$), of the spent fuel will not be recovered and becomes a waste material.

The long term radiological impact of the OTC can be controlled by a human-made system and natural barriers that should provide protection for as long as the life of the radiological source term they confine. The long time periods involved require a careful analysis of the confinement technology and of the long term consequences of conceivable scenarios.

At present, there is no worldwide agreement on the time intervals for the confinement of high level radioactive waste in a geologic repository. Periods of 1000, 10 000, 100 000 years or even longer have been considered, but no internationally accepted confinement period has been established.

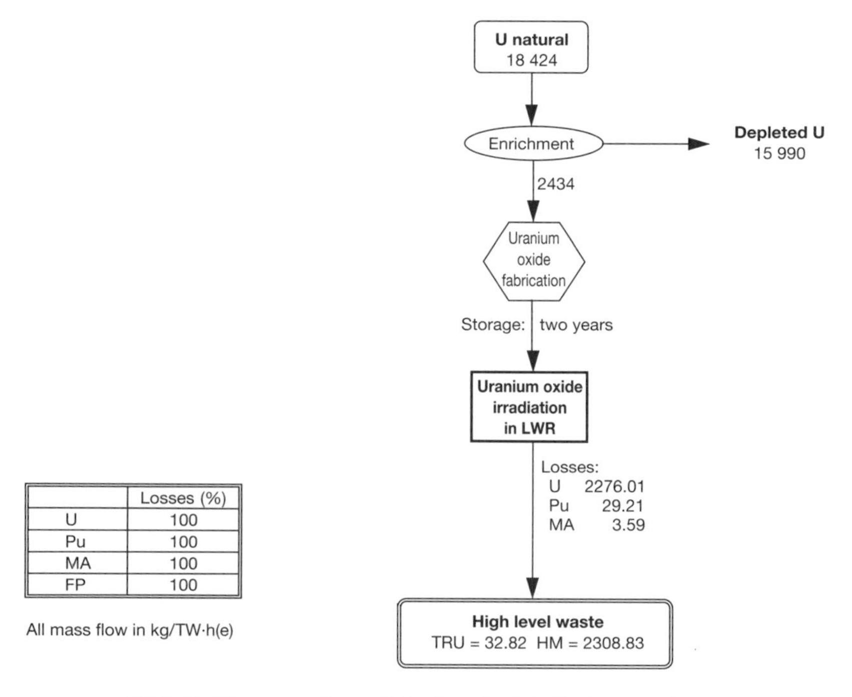


| | Losses (%) |
|---|---|
| U | 100 |
| Pu | 100 |
| MA | 100 |
| FP | 100 |

*FIG. 5. The once through fuel cycle. FP: fission products.*

### 4.1.2. Plutonium recycling in light water reactor MOX

Since natural uranium contains only 0.72% of the fissile $^{235}U$ isotope, the recycling of uranium and plutonium from spent fuel through the RFC scenario (Fig. 6) has been from the beginning of the nuclear era the standard scenario of nuclear energy production. There has, however, been reduced support for this approach in many States in recent years, owing to economic factors and particularly to proliferation concerns.

By processing according to this RFC scenario, the major fraction (~99.9%) of the uranium and plutonium streams is extracted and only a very minor fraction of the major actinides is transferred to the HLLW (and consequently to the vitrified HLW) and eventually to the geologic repository.

However, if public and/or political acceptance of very long term disposal of HLW cannot be obtained, the removal of MAs from high active residua or HLLW would be a technical solution that might reduce the residual radiotoxicity of the HLW. Moreover, with increasing burnup, the generation of MAs becomes more and more important. The addition of an MA partitioning

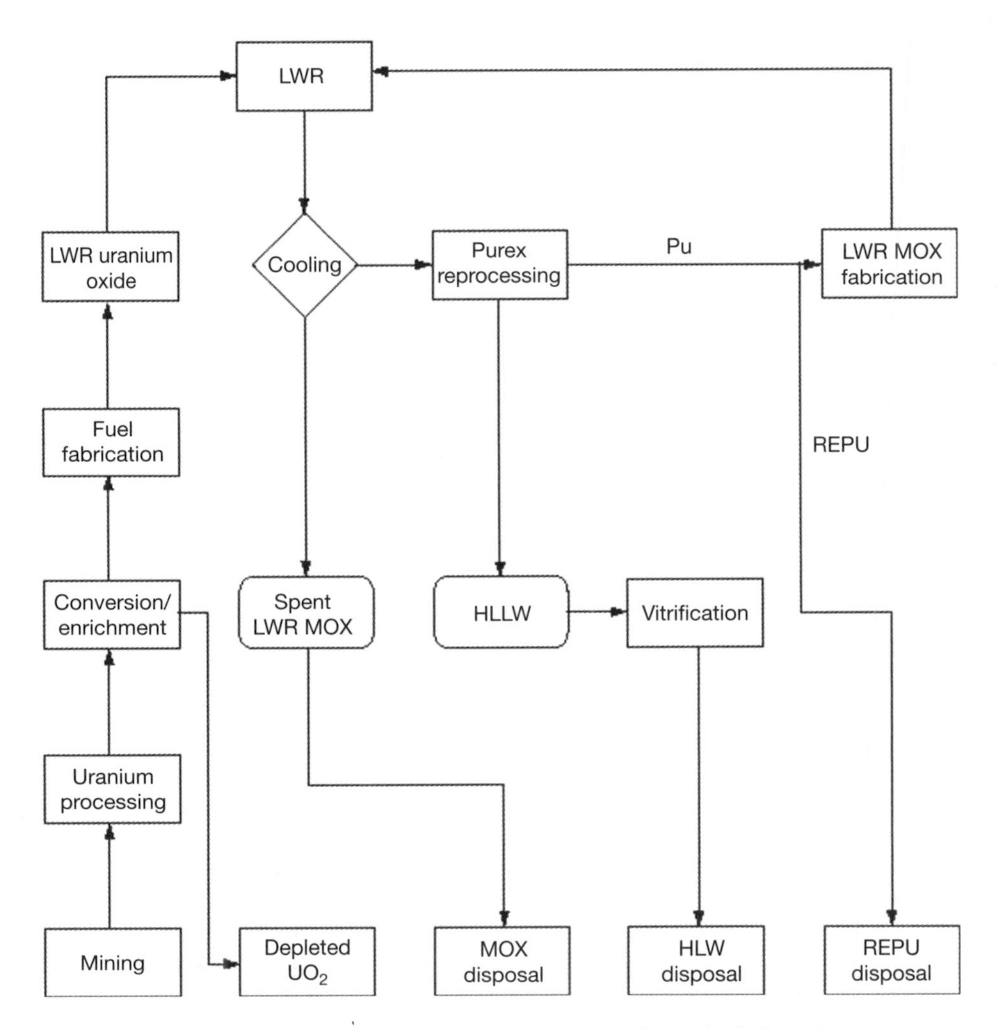


*FIG. 6. The LWR reprocessing fuel cycle. REPU: residual plutonium.*

module to the standard reprocessing plant would, in such a case, be the most obvious change to the current RFC. States with a reprocessing infrastructure (China, France, India, Japan, the Russian Federation and the UK) and their associated partners could in the medium term realize a partial partitioning scenario by which the HLW would be practically free from long lived TRUs.

However, the question arises of what to do with the recovered uranium, plutonium and MA fractions. Those States that chose to reprocess their spent fuel did this with the main purpose of recovering the major actinides (uranium and plutonium), to save on fresh uranium purchase (20%) and to use the residual fissile components of the spent fuel (c. 1% $^{235}U$, 1% Pu) corresponding to about 25% of the regular expense of the uranium enrichment step.

From the radiotoxic point of view the overall gain is rather limited, since only about 25% of the recycled plutonium is consumed and about 10% is transformed into a long term radiotoxic MA source term. Recycling of spent LWR fuel as MOX provides an overall mass reduction, with a factor of about five, but does not significantly reduce the total radiotoxicity. Double, or perhaps at the limit triple, recycling of LWR MOX is theoretically possible in LWRs if fresh plutonium is available, but in the short term the resulting radiotoxicity drastically increases throughout the subsequent recycling campaigns because of $^{244}Cm$ buildup [23]. This avoids reprocessing or fabrication of new fuels.

#### 4.1.3. Plutonium recycling in light water reactor MOX and fast reactor MOX

The stock of plutonium accumulated at the large European reprocessing plants that was intended to be used in liquid metal fast breeder reactors (LMFBRs) became redundant in a cheap uranium market economy. As a result, the use of separated plutonium achieved industrial significance in western Europe, where increasing quantities of $PuO_2$ were transformed into LWR MOX fuel and irradiated in specially licensed reactors in Belgium, France, Germany and Switzerland.

The reuse of plutonium is to a certain extent a first step in a global P&T scenario that has to be brought into a broader perspective of reuse of resources and reduction of the long lived waste produced during the nuclear age. However, if LWR MOX use is to continue for a long period of time there will be an accumulation of spent LWR MOX fuel, since there is at present little incentive to reprocess it.

This fuel offers, after two successive recyclings in an LWR, a degraded plutonium spectrum that makes it unsuitable for further thermal reactor operation. The only solution for this non-sustainable situation is the creation of a significant FR capacity, an option that was envisaged many years ago.

Reprocessing of LWR MOX fuel together with LWR $UO_2$ (in a ratio of 2:1) can be performed in the present large reprocessing plants. However, the reprocessing of a pure spent LWR MOX stream at an industrial throughput is beyond the capability of the present plants. A campaign for the reprocessing of 5 t of spent FR MOX fuel has been successfully carried out at the La Hague plant in France. The recovered plutonium can be transformed in FR MOX fuel in existing MOX fuel fabrication plants; Fig. 7 shows a flowsheet and mass inventories of such a cycle.

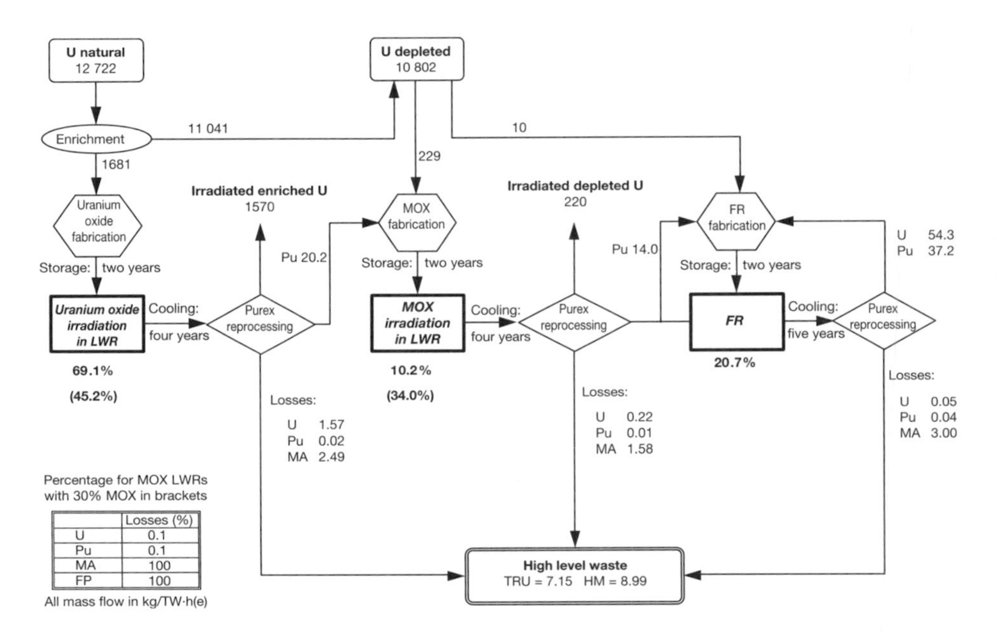


*FIG. 7. The LWR ($UO_2$ and MOX) FR fuel cycle with conventional and advanced Purex reprocessing.*

The real bottleneck of this scenario is the construction and financing of a sufficiently large FR fleet, which must be associated with an equivalent FR MOX reprocessing capacity. Based on flowsheet calculations [24] the FR fleet would gradually have to be upscaled towards a fraction of 36% FRs in order to achieve at equilibrium a zero buildup of the plutonium mass.

Multiple recycling by aqueous processing of this type of spent fuel cannot be accomplished without the introduction of fast centrifugal extractors in the first extraction cycle of the Purex process, and MAs would accumulate in the HLLW.

#### *4.1.3.1. Associated MOX fuel fabrication and refabrication problems*

A large amount of industrial experience has been gained in FR MOX fuel fabrication, since FR programmes have been undertaken in many nuclear States for several decades. The fabrication of FR MOX fuel with 15–25% plutonium has been realized routinely and on a commercial basis. However, the plutonium quality used for these purposes is derived from low burnup $UO_2$ fuel with low $^{238}Pu$ and $^{242}Pu$ contents. The burnup of spent LWR $UO_2$ and LWR MOX has, however, reached 50 GW·d/t HM. The isotopic composition of plutonium resulting from the reprocessing of such fuels is seriously degraded, with higher $^{238}$Pu and $^{242}$Pu levels and lower $^{239}$Pu and $^{241}$Pu concentrations.

Still higher plutonium concentrations are envisaged (up to 45%) in the use of advanced FBuRs (CAPRA). The recycling of fuels containing high $^{238}$Pu levels and limited amounts of MAs is still more difficult and requires the design and construction of remotely operated fuel fabrication plants.

For homogeneous recycling of MAs in FR MOX, admixtures of 2.5% $^{237}$Np and/or $^{241}$Am are currently being studied. Neptunium-237 is a pure alpha emitter (except for the small in-growth of $^{233}$Pa), and there is no major handling problem involved; however, the admixture of $^{241,243}$Am at the 2.5% level will induce a gamma field around the gloveboxes or hot cells. The major interfering radionuclide in FR MOX is $^{238}$Pu at the 3% level, which is a heat and neutron source.

FR MOX fuel fabrication with a 2.5% americium admixture will also be influenced by the degree of separation of the rare earths (strong gamma emitters) and by $^{244}$Cm, which will accompany the americium fraction when separated from HLLW. The presence of ~17% $^{244}$Cm in the americium and curium fraction of recycled FR MOX fuel will further increase the neutron emission up to $\sim 4.4 \times 10^{10}$ n/s per t HM.

The separation coefficients from rare earths and $^{244}$Cm required in order to permit industrial fuel fabrication operations will greatly depend on the permissible rare earth concentration in fresh FR MOX fuel and on the permissible $^{244}$Cm concentration in the fuel fabrication plant. The industrial processing of MAs in an extended reprocessing–partitioning option calls for the design and construction of dedicated MA fuel fabrication facilities in the vicinity of the major LWR $UO_2$ reprocessing plants.

It has to be remembered that specially equipped MA fuel fabrication laboratories, for example those of the ITU, are licensed to handle 100 g of $^{241}$Am and 5 g of $^{244}$Cm [25].

Heterogeneous recycling of MAs is a means of avoiding the dilution of troublesome radionuclides, for example $^{244}$Cm, throughout the fuel fabrication step and is carried out in small but dedicated and heavily shielded facilities. The waste handling resulting from the high specific activity of the MA radionuclides will strongly influence the generation of secondary waste with high alpha activity. The operation of hot cell type fuel fabrication plants is a new dimension in waste handling that will have to be assimilated to the present vitrification technology in terms of radiotoxic throughput.

The expected volumes of alpha active secondary waste call for new treatment technologies or for the integration of the secondary waste types in the mainstream of HLLW vitrification.

The development of specific MA waste treatment and conditioning methods using very sophisticated equipment as dedicated incinerators for alpha waste and production facilities for ceramic or cermet types of waste using, for example, hot isostatic pressing devices will be necessary if a full segregation between gamma active and long term alpha active waste types has to be accomplished.

### 4.1.4. Plutonium and minor actinide recycling in light water reactor $UO_2$ and MOX and in fast reactor burner transuranic elements

Consideration has been given in the recent past to using separated plutonium and MAs as a feedstock for FR fuel for the accelerated incineration of TRUs. This option requires special reactor core designs (IFR, CAPRA), advanced fuel types (oxides, metals, nitrides) and new or advanced reprocessing techniques (e.g. pyrochemical reprocessing), all of which require the development of adapted waste processing facilities. Figure 8 shows the LWR $UO_2$–Purex–TRU–IFR fuel cycle.

Co-processing of plutonium and MAs further increases the necessary (fissile) plutonium enrichment, the specific alpha activity of the fresh recycled fuel and the decay heat of the discharged spent fuel. The waste issues

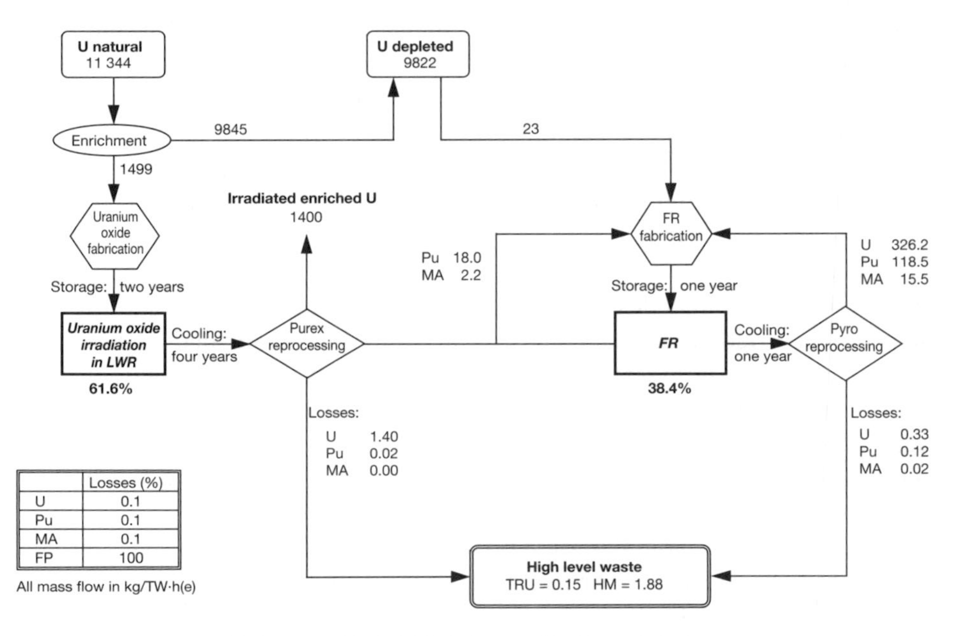


*FIG. 8. The LWR ($UO_2$)–Purex–TRU–IFR fuel cycle.*

associated with the use of oxide fuel are very similar to those to be encountered with FR MOX fuel discussed in Section 4.1.3. However, the use of new types of fuel (metal, nitride, etc.) calls for a specific discussion of the issues involved.

#### *4.1.4.1. Metal fuel fabrication for advanced liquid metal reactors and advanced fuels for burner reactors*

In the framework of the Integral Fast Reactor project [18], a specific fuel fabrication technology has been developed and tested on the cold (and hot) pilot scale. At the EBRII facility, metal fuel was recycled by casting a uranium–plutonium–zirconium alloy on the laboratory and hot pilot scale. It is obvious that these processes are still in the exploratory stage and cannot be considered as proven technology, but their potential should be investigated since metal alloy fuel permits very high burnups and has good material and neutronic characteristics for transmutation of TRUs. Uranium–plutonium–zirconium–MA alloy has been fabricated for property evaluation, and it is planned to irradiate it in an FR [26].

Attention has been drawn recently to the potential of nitride and carbide fuels [27] for FBuRs. Nitride TRU fuel containing macroscopic quantities of MAs can be produced by a combination of an internal gelation method and a carbothermic synthesis. These nitride fuels can be reprocessed by electrorefining methods similar to the technology developed for metal fuel.

Much technological experience has been accumulated during 30 years of R&D on fast breeder reactors worldwide, which can be transferred to FBuR technology.

Reprocessing of metal and nitride fuel relies on the use of pyrochemical processes and is followed by pyrometallurgical fuel fabrication for recycling in FRs or ADS systems.

### **4.1.5. Plutonium and minor actinide recycling in light water reactor $UO_2$ and MOX and accelerator driven system transuranic elements**

The reservation of certain governments to allow the separation of plutonium during reprocessing operations has led to the development in the USA of a scheme [28] in which the TRUs are kept together during the reprocessing and transferred to ADS facilities for transmutation–incineration. In this option 78% of the reactor fleet continues to be operated by LWRs, either LWR $UO_2$ or LWR MOX, depending on the availability of aqueous reprocessing. However, the remaining 22% of the electricity grid has to be fed by new reactor types, which are still in the design phase. This option (see Fig. 9) is obviously a long term solution of the TRU issue, relying principally on the availability of dependable ADS facilities capable of operating on a 24 h/d, 300 d/a basis.

In the supposition that such reactor fleets can be set up over the next 50 or 60 years, the question arises of what are the fuel cycle and waste management implications of this option:

(a) Aqueous reprocessing in its modified form (Urex) must be available for the majority of the spent fuel to eliminate uranium and some LLFPs ($^{129}$I, $^{99}$Tc). See Section 2.4.1 for an analysis.

(b) The separation of TRUs from the bulk of the fission products has to be performed by a pyrochemical process producing a concentrate of TRUs in metallic form ready to be recycled as solid fuel (metal or nitride) in the ADS reactor. Complete incineration of the TRU inventory implies the multiple recycling of TRUs in the ADS reactor and the effective separation of MAs from bulk rare earth fission products.

(c) Alternatively, an ADS reactor with a thermal neutron spectrum could be conceived with a molten salt core and continuous separation of TRUs from fission products. However, this option involves the development of a full pyrochemical reprocessing cycle with much improved TRU

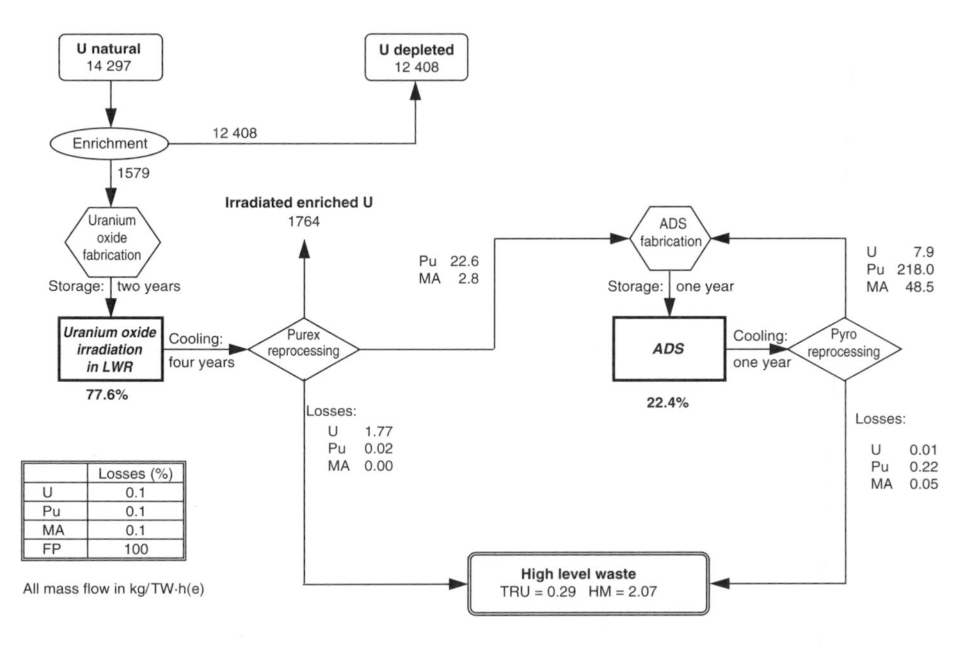


| | Losses (%) |
|---|---|
| U | 0.1 |
| Pu | 0.1 |
| MA | 0.1 |
| FP | 100 |

*FIG. 9. Advanced reactor and fuel cycle scenario with an LWR ($UO_2$)–Purex–ADS–TRU fuel cycle.*

separation yields. Experience was gained in the 1960s with a molten salt reactor, and a thorough review of the operation challenges and issues encountered with this reactor type should be made before making the proper choice.

(d) Waste management problems have not yet been fully assessed and only laboratory experience has been gained.

The geographic dispersion of the fuel cycle facilities with the LWR $UO_2$/ADS TRU option is one of the main factors for public acceptance. If 22% of the reactor fleet is converted into ADS facilities with pyrochemical reprocessing plants in the immediate vicinity of the ADS reactor, the aqueous reprocessing section (Urex) should be downscaled and associated with the location of the reactor and pyrochemical facilities. The multiplication of collocated reactor and recycling facilities throughout the territory of a State or continent representative of a 100 GW(e) reactor capacity is the major challenge for this option. However, if the sites for such activities can be selected and collocated with the waste disposal facilities, much less transport of spent fuel and waste concentrates will be required. The LWR $UO_2$/ADS TRU option favours a dispersed location of relatively small nuclear energy complexes coordinated with collocated waste disposal facilities.

#### 4.1.6. Plutonium and minor actinide recycling in a combined double strata strategy scenario

The double strata strategy initiated by the Japanese authorities in the late 1980s through the OMEGA programme [29] has received much international support as it allows the gradual development of a waste incineration scenario in conjunction with a straightforward nuclear electricity generating fleet. This option is very complex from the point of view of diversity in types of facility, but allows a gradual evolution from the present existing nuclear energy production fleet (LWR $UO_2$/LWR MOX) to a more efficient use of uranium by introducing FRs and by completing the scheme with the development and introduction of ADS facilities dedicated to the incineration of MAs (Fig. 10).

An equilibrium nuclear energy fleet under this scenario is made up of 65% LWR $UO_2$, ~10% LWR MOX, 19% FR (MOX or metal) and 6% ADS. Compared with the scenario described above, the difference lies in the dedication of the FR to plutonium recycling and in the reservation of the ADS capacity for MAs and residual plutonium destruction. This scenario continues to rely on aqueous reprocessing based on the well established Purex process and the newly developed aqueous partitioning methods for MAs. The recycling

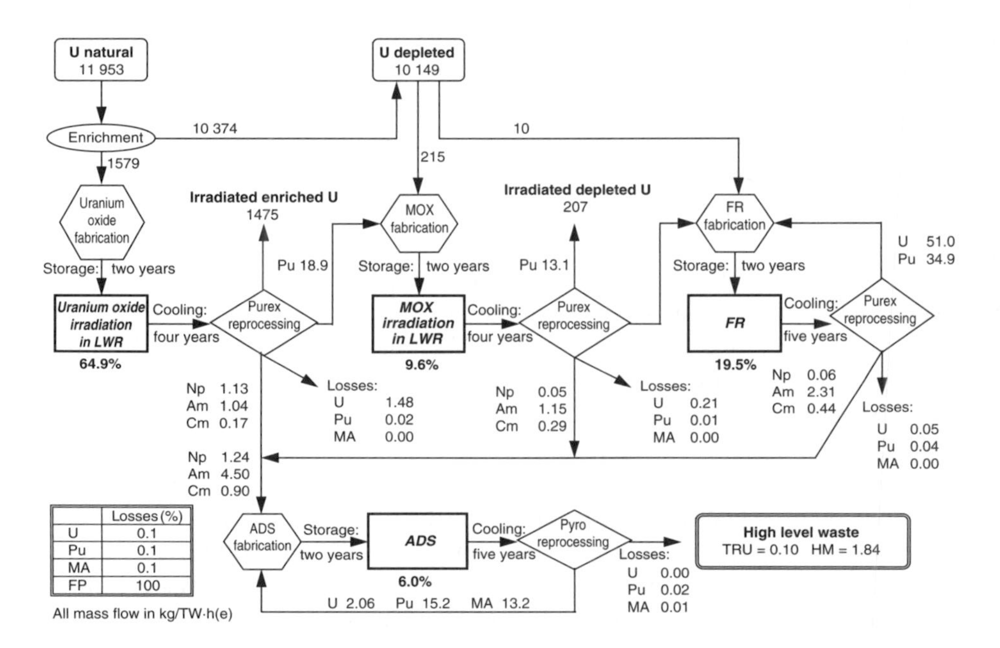


| | Losses (%) |
|---|---|
| U | 0.1 |
| Pu | 0.1 |
| MA | 0.1 |
| FP | 100 |



*FIG. 10. Double strata fuel cycle with an LWR ($UO_2$ and MOX) connected to FR MOX plutonium recycling and an ADS MA transmutation facility.*

of plutonium is possible for a long period of time (~50 years), since gradual buildup of the FR capacity can further rely on the reprocessing of mixed spent LWR $UO_2$/LWR MOX/FR MOX fuel in a ratio of 81%/12.1%/4.6%, respectively. The residual MA discharge resulting from the pyroprocessing corresponds to 3.4% and is to be treated in a centralized MA pyroprocessing facility.

By mixing 8 LWR $UO_2$ with 1 LWR MOX the total decay heat output only changes by a few per cent. Mixing the three types of spent reactor fuel (LWR $UO_2$/LWR MOX/FR MOX) increases measurably (~14%) the decay heat and the radiation damage of the extractant, but is still possible if fast centrifugal extractors are installed in the first Purex extraction cycle. The recycling of TRU fuel (MAs and residual plutonium) in the ADS reactor is limited to a relatively small throughput of 39 t HM/a for the whole 100 GW(e) reactor fleet. However, the decay heat of this fuel (~50% plutonium and 50% MAs) is very high and can only be treated by pyrochemical methods.

The consequences for the fuel cycle and waste management activities of the introduction of a double strata strategy at the industrial level are the following:

(a) The centralized reprocessing and waste management facilities existing, for example, in the European Union and Japan are fully capable of providing the necessary fuel cycle services for an integrated combined reactor fleet over several decades.
(b) The gradual increase of the $^{238}$Pu level in the recycled plutonium may call for more shielding and remote handling in the LWR MOX fuel fabrication plants.
(c) The most important change in the whole approach is the introduction of FRs in the nuclear electricity production scheme. However, there is ample time to adapt the FR technology to a safer operation, for example by replacing the sodium coolant by a less chemically reactive coolant (e.g. lead or lead–bismuth) or any other alternative low melting liquid metal.
(d) The advanced aqueous partitioning methods for MAs have to be implemented at the reprocessing sites and equipped with sufficient intermediate storage capacity in order to bridge the time span between current processing and the future transmutation–incineration era.
(e) The new reactor technology, specifically for ADSs, and the pyrochemical reprocessing technology for highly active TRU recycle streams are limited to a few per cent of the whole nuclear energy infrastructure. The gradual upscaling of the required ADS and pyrochemical facilities can easily be realized within the coming decades.
(f) The ADS reactor capacity will be installed preferably in the vicinity of existing reprocessing plants, which avoids the transport of highly active materials on public roads.

The implementation of MA partitioning technology can in principle produce an HLW stream without a significant level of long lived TRU contamination.

It is necessary to remark that the results mentioned above are computed at the theoretical maximum power level using IMFs. Complementary studies have been realized using breeder fuels such as $^{238}$U or $^{232}$Th [30–36].

### 4.1.7. Generation IV

The Generation IV International Forum (GIF) for the development by 2030 of the next generation of nuclear reactors and fuel cycle technologies was initiated by the USA in 2001 and now includes ten participating States (Argentina, Brazil, Canada, France, Japan, the Republic of Korea, South Africa, Switzerland, the UK and the USA). In 2002 six Generation IV systems having significant potential were selected by the GIF, with the help of leading international experts, from hundreds of proposals. Selection was focused on four main criteria:

(a) Sustainable nuclear energy: the optimal use of resources to minimize nuclear waste.
(b) Physical protection and proliferation resistance: this includes non-proliferation criteria and resistance to terrorism and theft.
(c) Safety and reliability.
(d) Competitive nuclear energy.

The next generation of nuclear energy systems is considered deployable by at least 2030, with some possibly available as early as 2020. The names for the reactors correspond to the cooling system, and include a sodium liquid metal cooled reactor (SFR), very high temperature reactor (VHTR), supercritical water cooled reactor (SCWR), lead alloy cooled reactor (LFR), gas cooled fast reactor (GFR) and molten salt reactor (MSR).

## 4.2. DUAL PURPOSE CONDITIONING FOR TRANSMUTATION AND/OR DISPOSAL

### 4.2.1. Fuel/target fabrication of minor actinides

Separated neptunium, americium and curium from the extended reprocessing and partitioning of a 100 GW(e)·a fleet would produce an annual output of:

(a) 1.6 t of neptunium as $NpO_2$;
(b) 1.6 t of americium as $AmO_2$;
(c) 0.2 t of curium as $CmO_x$.

The transmutation of such quantities can only be envisaged in several decades when innovative transmutation technologies have been developed at the industrial level. Two irradiation technologies can be used for this purpose: dedicated FRs and ADS reactors. The decision to build such facilities depends upon the specific energy policy of each State, regional federation or continent. Section 2.3 provides a general outlook on the waste management implications of such a strategy.

Taking into account the uncertainty about future decisions with regard to separated MAs and the time span involved, it would be appropriate to condition the separated radionuclides immediately after separation into a matrix that can either be used for irradiation in a transmutation facility or be transferred to a deep disposal site if the transmutation technology is not available.

The conventional irradiation matrices (oxides mixed with metal powder) cannot be used for disposal because of their solubility in deep aquifers and because the irradiation technology requires very specific physicochemical and nuclear characteristics for the design of irradiation targets.

Important progress has recently been made in the synthesis of crystalline matrices for the disposal of excess weapon plutonium. Some of these materials could be proposed for the conditioning of separated MAs.

#### *4.2.1.1. Zircon*

Zircon (zirconium silicate) [37–39] has been considered as a host phase for the disposal of weapon plutonium. The mineral exists in nature with a substantial uranium–thorium loading. Synthetic minerals with approximately 10% plutonium have been synthesized. Taking into account the isomorphous substitution of the americium, plutonium and neptunium for zirconium in the $ZrSiO_4$ matrix, decay processes of the actinides might not influence the long term holding capacity of the synthetic mineral. Studies on the effect of alpha radiation from $^{238}Pu$ have resulted in a comprehensive understanding of the amorphization yield and of the temperature zone within which it occurs. When neptunium is incorporated into zircon the alpha dose is very low and will not lead to significant amorphization, even after long periods of time. However, irradiation in a neutron flux with formation of $^{238}Pu$ will strongly accelerate the amorphization process.

During storage in natural media the thermal conductivity of the matrix is not a very important parameter, except for the storage of curium, but in irradiation processes it is a critical physical property.

The crystalline zircon material loaded with neptunium and/or americium has therefore to be diluted with zirconium metal powder in order to increase its thermal conductivity. Pellets with a cermet structure (80% zirconium, 20% zircon) are possibly the most adequate target form for irradiation and are acceptable for long term storage and disposal. Swelling of the structure up to 18% due to amorphization by alpha irradiation is the main disturbing phenomenon. During high burnup irradiation this phenomenon may be strongly amplified and therefore needs very thorough examination before this matrix is considered.

From the neutron economics point of view, the slightly higher microscopic neutron cross-section of the matrix makes this proposal somewhat less attractive than pure $ZrO_2$. However, the fabrication technique (powder mixing, pelletization, sintering) is much easier than the hot isostatic pressure technique for remote handling in hot cells.

*4.2.1.2. Zirconia inert matrix*

A highly radiation resistant material that shows promise for use as a durable storage material for plutonium is gadolinium zirconate. However, for irradiation purposes gadolinium should be substituted by a similar rare earth element with a very small neutron cross-section. Natural zirconium contains a small hafnium impurity with a high neutron cross-section, therefore nuclear grade zirconium should be used for this purpose.

The incorporation of plutonium or MAs in the crystal lattice of zirconium oxide depends on the degree of doping with trivalent rare earths, which creates vacancies in the lattice.

Development efforts should be started to determine the neptunium or americium and curium uptake within the crystal lattice. Isomorphic substitution of neptunium in its different valencies is very similar to that of plutonium.

*4.2.1.3. Rock-like yttria stabilized zirconia and spinel*

An inert matrix material of the zirconium type [40] has been prepared from aqueous solutions of zirconium nitrate, $Y_2O_3$, $Gd_2O_3$, aluminium nitrate and MgO, which was calcined at 1000 K and fired at 1670 K. The inert matrix was mixed with $PuO_2$ and pressed into pellets. The pellets were introduced into fuel pins and irradiated in the JRR-3M reactor of the Japan Atomic Energy Research Institute (JAERI). Post-irradiation examination by crystallographic methods showed the formation of different compounds but revealed mainly the occurrence of swelling and the release of fission products.

*4.2.1.4. Yttria stabilized zirconia based inert matrix fuel*

An IMF material [41] has been prepared by internal gelation of the oxyhydroxide phase from zirconium–yttrium–erbium nitrates and plutonium nitrate solution and a second preparation has been made using the attrition–milling of $ZrO_2$–$Y_2O_3$–$Er_2O_3$–$PuO_2$ followed by compaction and sintering. Owing to the addition of yttrium oxide, the thermal conductivity was shown to be lower than that of $UO_2$ and MOX. The material shows a very strong thermodynamic stability, and keeps it under irradiation conditions. Furthermore, this material looks very attractive for a ‘once through and then out’ irradiation strategy, since it remains very stable in reducing aquifer media.

#### 4.2.1.5. *Americium and curium incorporation into zirconia based materials*

Hot chemistry studies have been undertaken by the Commissariat à l'énergie atomique (CEA) to investigate the incorporation of americium and curium into yttria stabilized zirconia (YSZ) and into zirconate materials with a pyrochlore structure [42]. $AmO_2$–$ZrO_2$–$Y_2O_3$ displays a homogeneous ternary structure but suffers from sintering at high temperature. A better compound happens to be a zirconate compound with a pyrochlore structure, which can also incorporate americium as curium up to 30 mol. %. More elaborate studies have to be undertaken in order to determine whether such compounds could be used as an intermediate storage structure for separated MAs useful for irradiation or disposal.

### 4.2.2. Current research and development studies on inert matrix fuels

IMFs are foreseen to utilize plutonium in LWRs and to destroy it in a more effective way than MOX. After utilization in an LWR, the plutonium isotopic vector in the spent IMF foreseen for direct disposal will be devaluated far beyond standard spent fuel. Recent developments in IMFs have led to the selection of YSZ doped with erbia and plutonia at the Paul Scherrer Institute (PSI) in Switzerland [41], calcia stabilized zirconia doped with burnable poison and fissile material at the Agency for New Technologies, Energy and the Environment (ENEA) and the Politecnico di Milano in Italy and composites including YSZ doped with gadolinia and plutonia embedded in spinel ($MgAl_2O_4$) at JAERI [40]. See Annex II for more detailed analysis and references.

# 5. PARTITIONING

## 5.1. PARTITIONING OF MINOR ACTINIDES FROM AQUEOUS REPROCESSING STREAMS

As mentioned in Section 1, a very high separation efficiency is required to reduce the long term radiotoxicity of HLLW by a significant factor. Since the TRUs constitute the source term of the long term radiotoxicity, their removal from HLLW before vitrification is a necessary step in a partitioning strategy. Current aqueous reprocessing of spent fuel has a separation efficiency in the range of 99.8% to 99.9% for uranium and plutonium, and this figure can be improved to 99.9%. Based on this state of the art reprocessing technology, the goal of the MA partitioning step should be a separation efficiency of 99.9% (i.e. a decrease of the TRU content in HLLW by a factor of 1000). The MAs to be considered are neptunium, americium and curium, which are present in a strong nitric acid solution (~3M $HNO_3$) of the HAR stream.

### 5.1.1. Partitioning of neptunium

Recovery of $^{237}$Np from the uranium–plutonium product stream is technically possible in the Purex process. During current reprocessing operations, neptunium is partly discharged with the fission products into the HLLW and partly associated with the uranium, plutonium and neptunium stream in TBP. The purification of uranium and its quantitative separation from neptunium is achieved in the second extraction cycle of the Purex process. It would be advantageous to modify the parameters in the first extraction cycle in order to co-extract the three actinides (uranium, plutonium and neptunium) quantitatively and to recover the purified neptunium stream directly during the reprocessing. This will require an adaptation of the first extraction column and modification of the current uranium–plutonium separation flowsheet (Fig. 11).

Quantitative recovery of neptunium from dissolved spent fuel streams could be applied without major modifications in reprocessing plants.

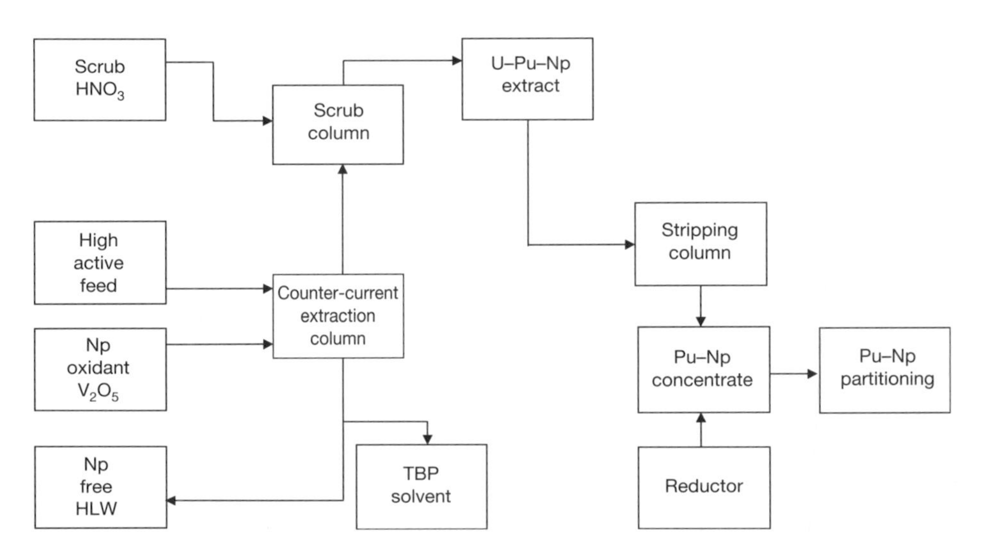


*FIG. 11. Purex flowsheet for neptunium separation.*

### 5.1.2. Partitioning of americium and curium from high level liquid waste resulting from spent light water reactor fuel

During conventional reprocessing operations, most of the MAs (neptunium, americium and curium) are transferred to the HLLW. Americium and curium (together with shorter lived TRUs: berkelium, californium, etc.) are quantitatively (>99.5%) transferred to HLLW. The partitioning of americium (plus curium) from HLLW is the first priority from the radiotoxic point of view and is also a prerequisite for a significant reduction of the (very) long term radiotoxicity due to $^{237}Np$. The separation of $^{241}Am$ obviously implies also the separation of the long lived $^{243}Am$, parent of $^{239}Pu$.

Partitioning of all MAs from HLLW is currently under investigation in many laboratories around the world (in China, France, India, Japan and some other States) and has been studied in the US national laboratories (Argonne National Laboratory, Oak Ridge National Laboratory and Hanford).

The americium and curium fraction contains all the rare earth elements, which are, in terms of quantity, about 10–20 times more important than actinides, depending on the burnup. At 45 GW·d/t HM the ratio is 16:1 (13.9 kg rare earth compared with 0.870 kg of americium and curium per t HM of spent fuel). Several processes have been studied and tested in hot facilities; among the most important are the TRUEX (Fig. 12), DIDPA, TRPO and DIAMEX (Fig. 13) processes for actinide–lanthanide group separation, coupled to the Cyanex 301, SANEX, ALINA and BTP (bis-triazinyl-1,2,4-pyridines) processes, which allow actinide–lanthanide separation.

The most important criterion to be used in ranking the different methods is the overall DF obtained during extraction of HLLW and its comparison with the required DFs in order to reach the 100 nCi level of alpha active radionuclides in the HLW. The highest DFs should be reached for $^{241}Am$ separation, namely $3.2 \times 10^4$ if immediate separation is scheduled. However, this option cannot yet be realized in industrial facilities, in which DFs of 1000 are the realistic limit. Another criterion is to minimize the feed conditioning in order to generate secondary waste during partitioning operations. Currently two candidate processes, namely TRUEX and DIAMEX, are being tested in hot cells in the European Union, India, Japan and the USA. Both processes have the potential to obtain DFs of $\sim 10^3$ for MA removal from acidic raffinate HLW. Based on recent hot cell tests carried out at the Joint Research Centre–ITU at Karlsruhe, the DIAMEX process represents the best compromise among the first series of methods. In laboratory conditions DFs of ~1000 have been obtained for the MAs from 3.5M acid concentrated HLW.

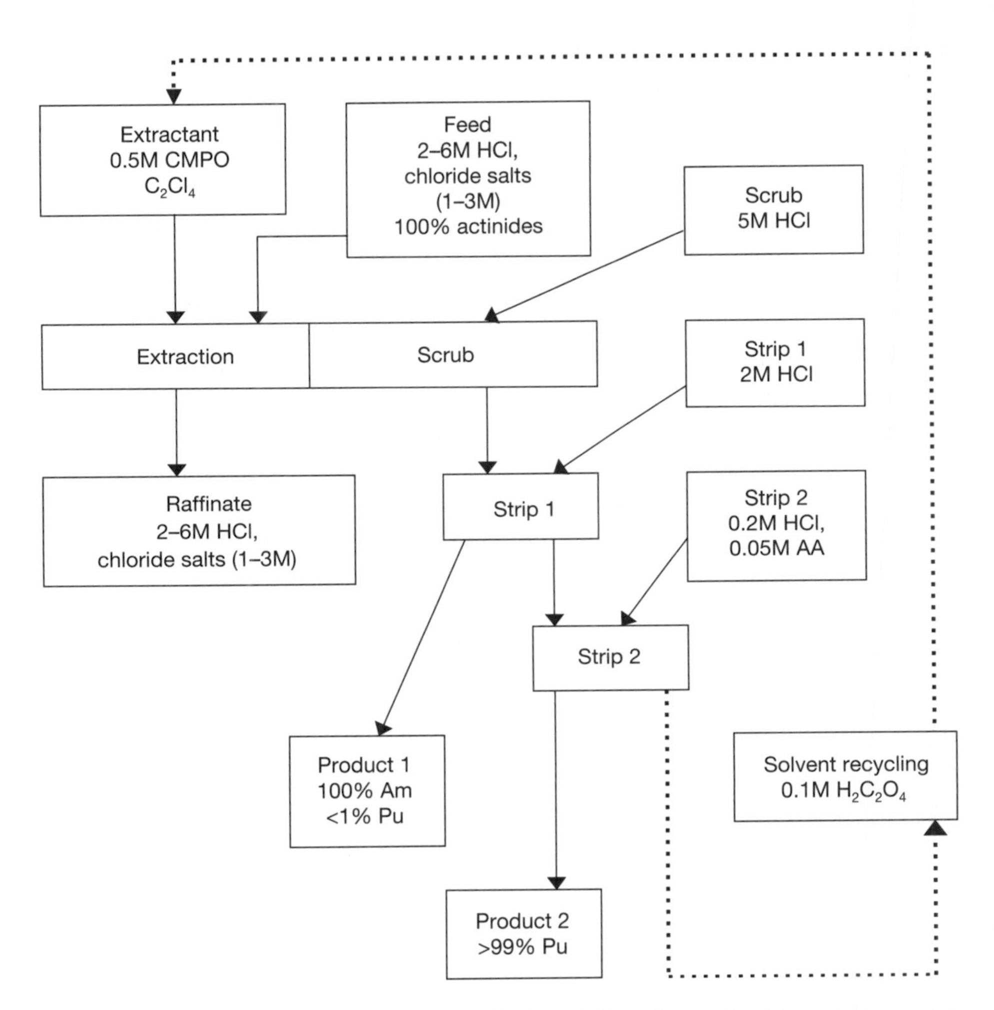


*FIG. 12. TRUEX flowsheet for removal of actinides from chloride solutions. AA: ascorbic acid. Solid lines represent aqueous streams. Dotted lines represent organic streams.*

Separation of actinides and lanthanides has been demonstrated with hot solutions using the BTP extractant and synthetic spiked solutions with the ALINA process based on the use of a new organosulphinic acid extractant, and seems to work properly in laboratory conditions. DFs of about 30 were obtained in 0.5M acid solution. One combined flowsheet is shown as an example in Fig. 13. In order to obtain an MA fraction with 90% purity, an actinide–lanthanide DF of more than 100 is required for the rare earth fraction. A 99% purity involves an actinide–lanthanide separation factor of more than 1000.

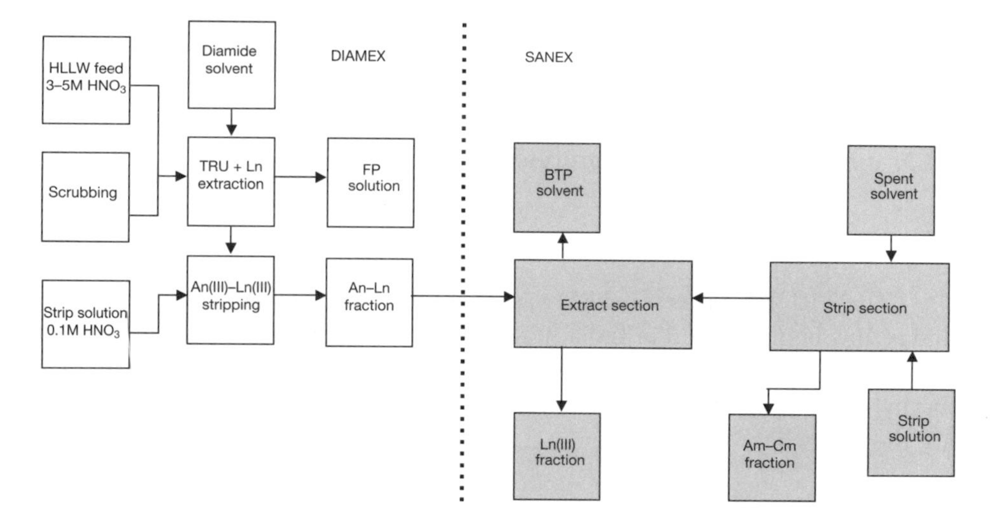


*FIG. 13. DIAMEX–SANEX conceptual flowsheet.*

The TRUEX process is also very effective for alpha decontamination of medium level and non-heating high active waste (HAW) streams.

Except for the US ex-military facilities, in which kilogram scale separations have been performed, the present research facilities, in the European Union, have strong limitations with regard to the quantities of MAs that can be handled in shielded facilities; for example, the new MA laboratory of the Joint Research Centre–ITU has an authorization for a maximum of 150 g of $^{241}$Am and 5 g of $^{244}$Cm.

As a nuclear power plant fleet of 100 GW(e) produces annually about 1600 kg each of neptunium and americium and curium, a big chemical engineering effort will be needed to upscale the laboratory methods to a pilot scale, and subsequently to an industrial prototype scale, in future advanced reprocessing plants in order to include MA separation rigs from the design phase on.

Looking at the existing situation in the European Union we can take an industrial RFC for granted. There is sufficient reprocessing capacity (La Hague, Sellafield) to cover the European and some overseas needs for the next 20–30 years. The first steps to implement the advanced fuel cycles are the installation of the separation facilities for MAs from HLLW and the conditioning of these radionuclides for intermediate storage or as a potential target material for transmutation.

HLW free of actinides, or at least HLW depleted of actinides, could be produced by vitrification plants and stored for cooling in surface facilities followed by geological disposal. There are no objective arguments to oppose geological disposal of such a waste stream, which decays with more than four orders of magnitude over 500 years.

### 5.1.3. Status of partitioning processes

#### *5.1.3.1. TRUEX process*

The TRUEX process, which was developed in the USA in the 1980s [43] and is now being studied in India, Italy, Japan, the Russian Federation and the USA, is based on the use of the CMPO (octyl-phenyl-di-isoburyl-carbamoyl-methyl-phosphine-oxide) extractant. The advantages of the TRUEX process are the following:

(a) It can extract actinide (and lanthanide) salts from acidic feeds;
(b) Its efficiency has been demonstrated with genuine HAW;
(c) A large amount of experience has been gained worldwide.

The main drawbacks of the TRUEX process are the following:

(i) The necessity to use a large concentration of TBP as a solvent modifier added to the solvent to prevent third phase formation;
(ii) There is stripping of metal ions, which is not efficient;
(iii) The delicate solvent cleanup.

#### *5.1.3.2. DIAMEX process*

The DIAMEX process was developed in France [44] and is now under research in France, Germany, Italy, India, Japan and the USA; it is based on the use of a malonamide extractant. The main aspects of the process are the following:

(a) Actinide (and lanthanide) salts can be extracted from undiluted acidic feeds;
(b) Its efficiency has been demonstrated on genuine HAW (in France);
(c) No secondary solid wastes are generated, owing to the CHON (use of extractants with radicals having only carbon, nitrogen, oxygen and hydrogen) character of the malonamide extractant.

A process based on a new type of diamide, a diglycolamide, which is a terdendate ligand having better affinity for An(III) than the malonamide extractant, is under development in Japan [45].

#### *5.1.3.3. TRPO process*

The TRPO process, which was developed in China [46] and is now under research in China and India, is based on the use of a mixture of tri-alkyl phosphine oxides ($R_3P(O)$, where R = alkyl groups) as an extractant. This process has been tested successfully in China with genuine HAW. Its main drawbacks concern the necessity to reduce the feed acidity and to use a concentrated nitric acid solution for An(III) + Ln(III) stripping, which complicates the subsequent An(III)–Ln(III) partitioning step.

TRPO work carried out in India [47] used a mixture of 30% Cyanex 923 and 20% TBP on diluted (1M $HNO_3$) synthetic HLW solutions. Quantitative Am(III) recovery was demonstrated in five to six stages.

#### *5.1.3.4. An(III)–Ln(III) separation*

(a) TALSPEAK and CTH processes

The TALSPEAK process, which was developed in the USA [48] in the 1960s and then adapted (CTH process) in Sweden [49], can be considered as the reference process for An(III)–Ln(III) group separation. It is based on the use of HDEHP (di-2-ethyl-hexyl-phosphoric acid) as extractant and DTPA (diethylene-triamine-penta-acetic acid) as the selective An(III) complexing agent. The An(III)–Ln(III) separation is performed by the selective stripping of An(III) from the HDEHP solvent loaded with a mixture of An(III) + Ln(III) under the action of an aqueous solution containing DTPA and a hydroxocarboxylic acid such as lactic, glycolic or citric acid. The advantages of this process are the large experience gained worldwide and its good efficiency. Among the main drawbacks are the necessity to adjust the pH of the feed, the limited solvent loading of metal ions and the difficult solvent cleanup. HEHEPA (hexylethyl hexylethyl phosphonic acid) is also reported (in India) to give good actinide–lanthanide separation factors. The advantage of this process is that the loaded organic can be stripped using dilute nitric acid alone. The drawbacks are the same as for HDEHP.

(b) SANEX concept (acidic S bearing extractants)

The Cyanex 301 process is under research in China, Germany and India; the extractant consists of a dialkyldithiophosphinic acid ($R_2$–PSSH, where R = an alkyl group). Its use for An(III)–Ln(III) was first proposed in China [50] in 1995. The main interest of the process lies in its good efficiency for An(III)–Ln(III) separation and in the fact that the process has been tested with genuine An(III) + Ln(III) mixtures. Nevertheless, for an efficient use of this process the feed solution should be adjusted to pH3 to pH5, which is not easy to carry out industrially. Moreover, the solvent cleanup is also a weak point.

Recent work carried out in India [51] has shown that the synergistic mixture of Cyanex 301 and 2,2′bipyridyl is particularly promising in view of the exceptionally high separation factor between Am(III) and Eu(III).

To cope with the main drawbacks of the Cyanex 301 process mentioned above, it was proposed to use a synergistic mixture made of bis(chlorophenyl)dithio-phosphinic acid and tri-n-octylphosphine oxide to perform the An(III)–Ln(III) group separation [52]. Even though the separation factors between An(III) and Ln(III) are less than those observed with Cyanex 301, the concentration of nitric acid in the feed can be as high as 1.5 mol/L, which makes the ALINA process more attractive than the Cyanex 301 process. The ALINA process (Fig. 14) has been successfully tested with genuine waste. The possible drawbacks of this process are that the solvent cleanup process is not yet defined and the generation of phosphorus, sulphur and chlorine bearing waste (from the degraded extractants), which needs to be managed.

(c) SANEX concept (neutral N bearing extractants)

After the discovery in Germany of the astonishing properties of BTPs for An(III)–Ln(III) separation [53], a process was developed and tested in the framework of the European PARTNEW project [54]. Successful hot tests have been achieved both at CEA Marcoule and the ITU using n-propyl-BTP. A good efficiency for the BTP process was obtained. It should be mentioned that the feed of the n-propyl-BTP process can be acidic (1M/L $HNO_3$). This process is very promising and may perhaps lead to the direct separation of americium and curium from acidic HLW without prior group separation of the rare earth–americium–curium fraction.

Nevertheless, it was shown at the CEA during a hot test that the n-propyl-BTP is not sufficiently stable against hydrolysis and radiolysis to be proposed as an industrial extractant. Research is currently under way in the European Union to design a more robust BTP extractant.

A synergistic mixture made of ter-dendate N-ligand, 2-(3,5,5-trimethyl-hexanoylamino) 4,6-di-(pyridin-2-yl)-1,3,5-triazine and octanoic acid has been

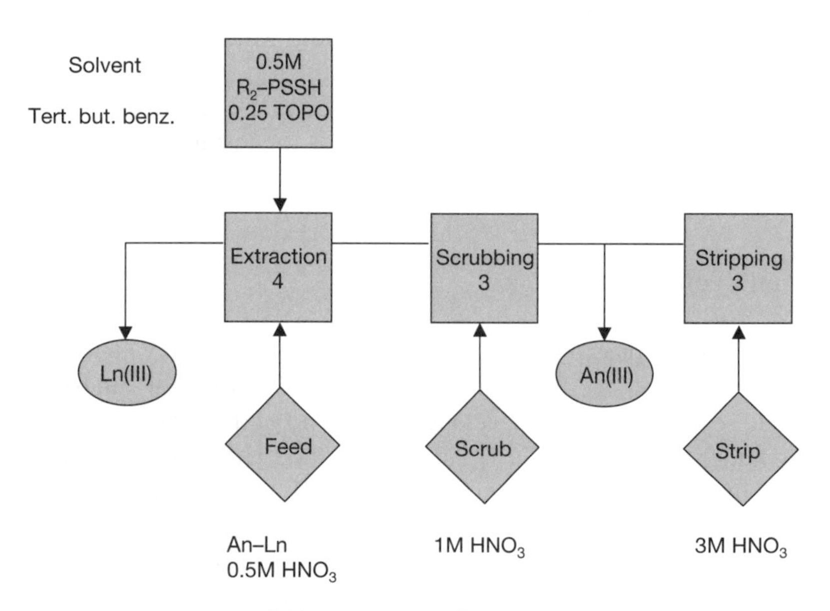


*FIG. 14. ALINA flowsheet.*

developed in France [55]. A process flowsheet has been defined and successfully tested with genuine effluent with good efficiency. The main drawbacks of this process are the required pH adjustment of the feed and that the management of secondary waste is not yet defined.

### *5.1.3.5. Americium and curium separation*

For this step, processes based on the selective oxidation of americium at the +VI or +V oxidation states have been developed, the curium remaining unchanged as Cm(III), allowing simple americium and curium separation processes to be defined.

(a) SESAME process

In strong oxidizing conditions, americium can be oxidized from Am(III) to Am(VI). This can be done, for example, by electrolysis in the presence of heteropolyanions acting as a catalyst. The generated Am(VI) can be separated from Cm(III) by extraction, for example by TBP. This is the principle of the SESAME process [55], which is under development in France. In Japan, oxidation of Am to Am(VI) is achieved by the use of ammonium persulphate; Am(VI) is then extracted by TBP. The SESAME process exhibits a great efficiency for americium and curium separation. Much experience has been obtained in France at the pilot scale over the past 20 years with a SESAME-like process (kilogram amounts of $^{241}Am$ were purified). Nevertheless, the industrialization of the process faces difficulties, such as the instability of Am(VI), the difficulty of developing a multistage process and the generation of secondary solid waste (made of heteropoly acid constituents).

(b) Am(V) precipitation

The selective precipitation of double carbonate of Am(V) and potassium is one of the oldest methods for americium and curium or americium and lanthanide separation; it was developed at the end of the 1960s in the USA and is today under development in Japan. This method requires the use of a 2M/L $K_2CO_3$ solution in which the mixture of Am(III) and Cm(III) is dissolved. After chemical or electrochemical oxidation of Am(III) to Am(V), Am(V) precipitates from the solution as the solid crystalline $K_5AmO_2(CO_3)_3$ $nH_2O$, while Cm(III) remains in solution. After filtration, americium is separated from curium. This process is simple, selective for americium and has been used worldwide. The main process drawbacks are the americium losses with curium, which are not low, the fact that it exists only as a single stage process and that large amounts of secondary waste are generated.

### 5.1.4. Comparative testing of advanced aqueous processes

Advanced aqueous processes (TRUEX, DIDPA, TRPO and DIAMEX) have been investigated and compared [56]. The common feature of the different approaches consists of a post-processing of the HAW with the objective of extracting the MAs.

For all four extractants, continuous counter-current extraction tests have been carried out in a centrifugal extractor battery. The extractants investigated have reasonably good extraction properties and should allow achievement of the very high DFs necessary for an effective P&T strategy. The most efficient stripping and the highest recovery rates are achieved with the DIAMEX and TRPO processes. With the DIAMEX process, in contrast to the TRPO process, no acidity adjustment (denitration) of the feed solution is needed. The DIAMEX process therefore represents the best compromise of all four processes studied, and shows good extraction and excellent stripping properties. The separation efficiencies are shown in Table 1.

The separation efficiency of liquid extraction processes is close to 99.9% for uranium and plutonium. It is expected that similar efficiencies could be obtained in the MA separation schemes for neptunium and americium and

TABLE 1. SEPARATION EFFICIENCIES FOR VARIOUS ACTINIDES AND FISSION PRODUCTS IN DIFFERENT CHEMICAL PROCESSES

| | Purex (industry) | Advanced aqueous reprocessing[a] (laboratory) | Pyrochemistry (laboratory) |
|---|---|---|---|
| Uranium | 99.9% | 99.9% | 99.9% (prototype) |
| Plutonium | 99.8% | 99.9% | 99.9% (prototype) |
| Neptunium | 95% | 99.9% | 99.9% |
| Americium | — | 99.9% | 99.9% |
| Curium | — | 99.3% | ? |
| Lanthanides in MAs | — | ≲5% | <10% |
| Caesium-135,137[b] | — | 99.9% | — |
| Technetium-99 | — | ~80% | — |
| Iodine-129 | 98% | 99.9% | — |

[a] Purex, DIAMEX and SANEX.
[b] Calixarenes.

curium. However, the americium and curium fraction is always accompanied by the bulk of the lanthanides. These have to be separated from americium and curium, but the efficiencies are much lower than 99.9%; between 5–10% of lanthanides in separated actinides may be expected. Technetium in liquid form ($TcO_4^-$) contains 80% of the total technetium inventory; the residual quantity remains as insoluble residues. Iodine can be efficiently separated (>99%) from the dissolver off-gases.

Unfortunately, none of these processes allow the separation of lanthanide fission products from MAs. For lanthanide separation from MAs, a two step partitioning process is required in which the aqueous lanthanide–MA fraction generated from the processes mentioned above (i.e. TRUEX, DIDPA, TRPO and DIAMEX) is subjected to the SANEX process, in which the MAs are selectively extracted from the lanthanide–MA fraction. The performance of several solvents has been compared and excellent results obtained for the BTP process [53] using the n-propyl-bis-triazinylpyridine molecule. The experiment, carried out in a centrifugal continuous counter-current set-up, achieved an MA–lanthanide separation with an efficient scrubbing of lanthanides and produced an MA fraction almost free of lanthanides. MA extraction and back-extraction were efficient and a reasonably good recovery of americium (>99%)

was achieved. Nevertheless, this process scheme has still to be improved to increase the recovery of curium (at present 97.6%).

By means of these tests it could be demonstrated that an efficient separation of MAs from genuine spent fuel is possible in a three step process (Purex, DIAMEX, BTP).

For transmutation in a fast neutron flux one would not need such high purification, since enhanced parasitic neutron capture of certain fission products does not exist in a fast neutron spectrum.

### 5.1.5. Conditioning of separated minor actinides and fabrication of irradiation targets

The potentially separated MAs could temporarily be stored as pure oxides, but since the critical mass of americium is 34 kg and that of neptunium is 55 kg, effective safety and security precautions need to be taken in order to safeguard the separated radionuclides during their storage period (see Section 3.4).

For waste management purposes the separated neptunium and americium and curium could preferably be mixed with a very insoluble matrix of zirconolite, hollandite and perowskite known as Synroc; up to 30 wt% plutonium and/or MAs can be immobilized this way. Once in the embedded form, retrieval of the radionuclides from the matrix is very difficult. Their solubility in geologic fluids is several orders of magnitude lower ($10^{-6}$ $g \cdot m^{-2} \cdot d^{-1}$) than that of conventionally vitrified waste. Since the leach rate and the solubility in groundwater determine the ultimate radiological risk, such a procedure would sharply decrease the long term risk of a repository in comparison with conventional vitrification of HLW. However, the criticality issue involved when large amounts of MAs are transferred to a repository remains to be addressed and calls for additional study. Mixing of conditioned MA fractions with very refractory neutron absorbing materials, for example boron carbide ($B_4C$) or rare earth mixtures, is a possible approach.

The second option is the transformation of the MA fraction into a ceramic irradiation target material ($Al_2O_3$, $MgAl_2O_4$, etc.) and its storage in critically safe fuel type storage racks until the transmutation technology becomes operational.

Management of the mixed MA fraction requires a very precisely determined strategy of what the fate of the separated radionuclides will be in a time frame of 100 years prior to their transfer to a geologic repository. If transmutation of MAs is chosen as an option before disposal, the induced heat associated with the formation of $^{238}Pu$ and $^{244}Cm$ will determine the necessary cooling period before transfer to a repository. Within this time frame it would be necessary to keep a type of TRU recycling technology open in order to reclaim or recycle unwanted waste streams.

New facilities will need to be designed and constructed for the conditioning of fuels or targets:

(a) Hot pressing or high isostatic pressing at elevated temperatures needs to be designed for use with highly alpha active powders emitting non-negligible neutron irradiation. Since 20–30 wt% alpha active material can be taken up in target compounds, facilities with a gross annual throughput of up to 5–8 t of both neptunium and americium and curium compound need to be made available for a 100 GW(e)·a spent fuel output. For a mixture of MAs, a highly shielded and remote handled (and serviced) fuel/target processing facility with a throughput of up to 10–16 t of MA compound needs to be constructed and operated over several decades.

(b) For the fabrication of americium and curium irradiation fuels or targets, new technologies, for example sol gel fabrication or inert matrix pellet impregnation (INRAM (infiltration of radioactive material)) are envisaged. The first technique can be used for medium active alpha emitters, but is probably not suited to very hot material (curium). The second technique is under laboratory investigation. An industrial pilot facility for up to 5–8 t of MA compound per year would be required to handle the annual output.

(c) For the fabrication of neptunium-containing fuel assemblies or target pins, conventional MOX fuel fabrication techniques can be employed. As the neptunium content of FR fuel may not exceed 2.5%, a MOX fabrication capacity of 64 t uranium–plutonium–neptunium fuel would need to be reserved for that purpose. If thermal irradiation of neptunium is proposed, the resulting $^{238}$Pu concentration in the irradiated target becomes the limiting factor in the neptunium enrichment of the initial uranium–plutonium–neptunium fuel.

It may be assumed that within the period 2015–2040 one or several of the above fuel or target fabrication technologies could become operational at the industrial level. Since the methods for waste conditioning and fuel fabrication are more or less similar, both options can be utilized from the outset, with preferably a dual option being realized (i.e. conditioning of MAs as potential irradiation targets).

In the medium term, only thermal reactors and particularly LWRs are available for irradiation of MA loaded fuels or targets. Fabrication of irradiation targets with industrially representative quantities of MAs is difficult to accomplish even in pilot scale hot cell facilities because of health physics implications for the operating crews in the plant.

The presence of large quantities of $^{241}$Am accompanied by 1–10% rare earth will require fully gamma shielded and remotely operated fabrication facilities. The presence of 5% $^{244}$Cm in an $^{241,243}$Am target will amplify the degree of technical complexity, due to the additional neutron shielding resulting from the spontaneous fission rate and from the $\alpha$–n reaction in oxide type isotopic targets. Extensive experience has been gained in the production of isotopic heat sources, but the present radiological context and the as low as reasonably achievable limitations to be expected from regulatory bodies on industrial activities render the recycling of MAs very different from in the past for military and space applications.

Conditioning of pure MA fractions with formation of very insoluble and thermodynamically stable compounds is a positive contribution to long term migration risk reduction. Since the heat output has not been modified from its level in spent fuel, having MAs in individual waste containers among the HLW canisters in underground facilities does not influence the overall repository design. However, it may have an effect on the local activation of structural and backfill materials.

The presence of conditioned MA concentrate in a repository reduces its dispersion throughout the spent fuel or HLW canisters and decreases the contamination risk in the event of accidental human intrusion. However, it does not fully eliminate this hazard.

Transmuted MA targets have to be kept in surface storage for extended periods of time, owing to the increased heat output resulting from the generation of $^{238}$Pu and $^{244}$Cm in the targets. The radiotoxicity of these targets is extremely high during a decay period that ranges from 200 years for spent americium targets to 800 years for neptunium targets.

## 5.2. PYROCHEMICAL REPROCESSING

For several decades pyrochemical and pyrometallurgical technologies have been investigated for multiple recycling of the very hot nuclear materials resulting from FR irradiation. Initial progress was achieved in the USA during the IFR project [57] and later in the Russian Federation [58]. Basic but important activity has been started on pyrochemical process development by installing a hot cell caisson in the CRIEPI–ITU facility at Karlsruhe.

Recently the Accelerator Transmutation of Waste project renewed interest in this technology and initiated the Roadmap project, which reviewed and updated the current state of the art [59]. This project has been transformed into a new proposal [4] as a follow-up to the Roadmap project.

A key requirement of the US Department of Energy proposal is that pure plutonium may not be separated or handled. The process treats all the transuranic radionuclides as a homogeneous group of elements that has to be treated simultaneously during chemical processing and the transmutation phase.

### 5.2.1. Principles of pyroprocessing

Originally, both aqueous and pyroprocessing were applied for military purposes: the Purex process to extract pure enough plutonium, the electrorefining process to clean plutonium metal from in-grown americium. This latter process was extended to reprocess metallic spent fuel from the EBRII reactor at Argonne National Laboratory–West [57]. In contrast to conventional aqueous processing, pyroprocessing does not involve dissolving spent fuel in an acid solution. Rather, the fuel is chopped and suspended in baskets in a molten salt bath in which an electric current is flowing. Most of the spent fuel constituents, including uranium, TRUs and fission products, dissolve into the salt. Whereas most of the fission products remain in the salt, uranium and TRUs are removed from the salt through deposition on different cathodes.

A demonstration of dry reprocessing of metallic uranium–plutonium–zirconium based MA fuels is being carried out within the framework of a study [60, 61]; a schematic diagram of the electrorefining is shown in Fig. 15. During

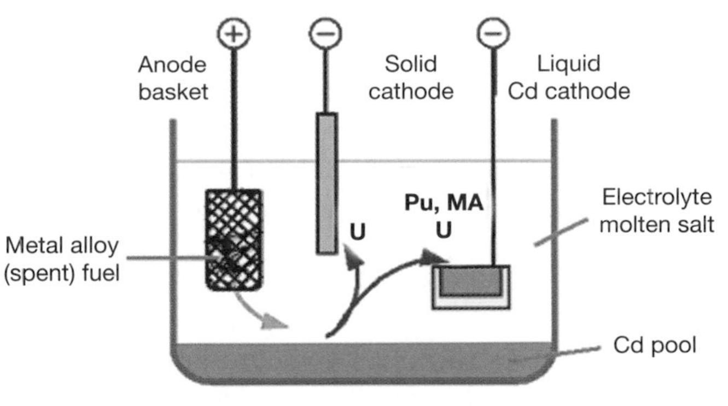


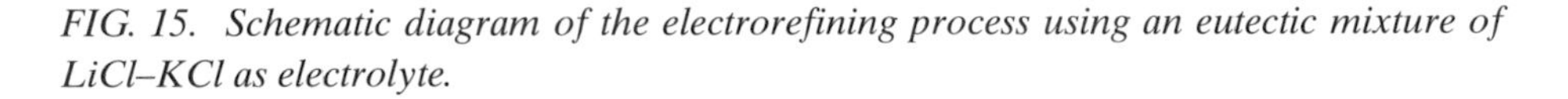

*FIG. 15. Schematic diagram of the electrorefining process using an eutectic mixture of LiCl–KCl as electrolyte.*

the process, metal alloy fuel in the anode basket is anodically dissolved (oxidized) into the molten salt electrolyte. At the same time, uranium metal is deposited (reduced) onto the solid cathode. After uranium recovery, the solid cathode is replaced by a liquid cadmium cathode into which a mixture of plutonium, uranium, MAs and a small amount of lanthanide fission products is recovered when the electrorefining is continued. Fission products, which have low redox potential, are not dissolved and remain in the anode basket or fall down into the bottom cadmium pool. More reactive fission products remain dissolved in the salt phase. Fission products less reactive than actinides are not dissolved and remain in the anode basket.

A successful installation of electrorefining equipment in a new experimental set-up has been achieved by CRIEPI and the ITU. Preliminary electrorefining tests on pure uranium and plutonium metals were carried out using a solid cathode and a liquid cadmium cathode to prove the operational capability of the facility.

For the reprocessing of (uranium, plutonium, zirconium, MA) fuels, uranium is electrodeposited on the solid cathode. In the next step plutonium is co-deposited together with MAs into a liquid cadmium cathode. The electrodeposition was carried out at a constant current of 500 mA for 4 h. The ingot is covered by solid salt; a blue colour comes from the $PuCl_3$ dissolved in the LiCl–KCl eutectic salt mixture [62].

After optimization of the TRU recovery from unirradiated (uranium, plutonium, zirconium, MA) metal fuels, reprocessing of irradiated fuel is planned for the coming years. The experimental facility will be installed in the ITU's hot cell laboratory under a contract with CRIEPI. The feasibility of electrodeposition of uranium on a solid cathode and of plutonium and americium in a liquid cadmium cathode could thus be demonstrated. Reductive multistage extraction is another potential method to recover actinides or to separate actinides from lanthanides in a molten metal (cadmium or bismuth) system. Distribution coefficients between molten chloride salt and cadmium–bismuth have been measured. Recovery rates by multistage reductive extraction obtained in a laboratory facility are very encouraging: 99.7% for plutonium, 99.8% for neptunium and 99.4% for americium

### 5.2.2. Pyroprocessing in the USA

In the USA, LWR $UO_2$ spent fuel would be the material subjected to the conventional chop and leach process in order to eliminate by aqueous extraction (the Urex process (Fig. 16)) the bulk of the uranium, technetium and iodine from the dissolved fuel mixture [4, 59].

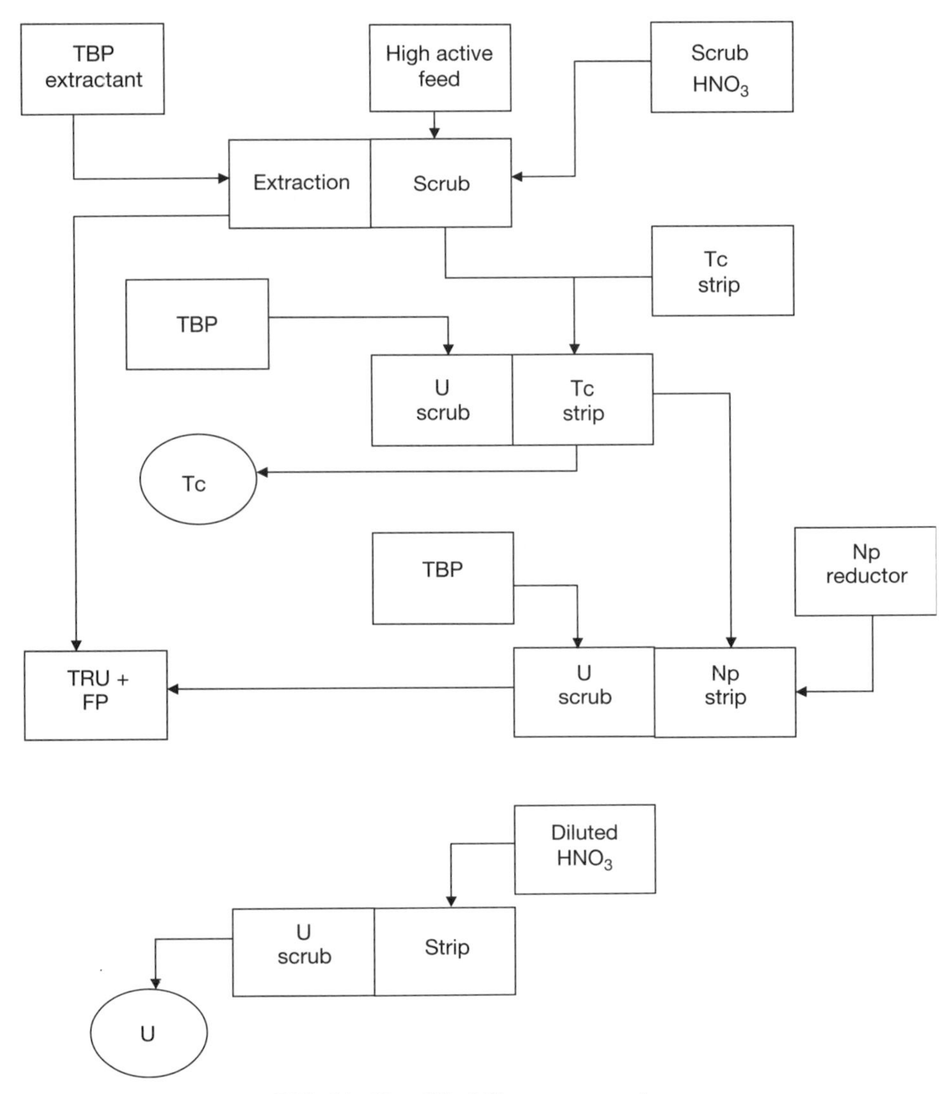


*FIG. 16. Simplified Urex process scheme.*

The raffinate of that extraction containing the fission products and the TRUs (~6% of the initial mass) is calcined and transferred to the pyrochemical section of the plant. The calcined oxide mixture of fission products and TRUs is submitted to a lithium reduction step with Li–LiCl in a furnace at 650°C. The alkaline fission products caesium and strontium remain in the salt mixture and are treated as waste. The TRUs, rare earths and residual metals (zirconium, molybdenum, etc.) are treated in an electrorefining furnace at 500°C. The electric potential of the cadmium electrode is selected at a potential where the rare earths do not yet deposit. The metal concentrate of the TRUs is mixed with zirconium metal to produce by casting a TRU–zirconium metal alloy fuel. Figure 17 shows a general flowsheet of the pyrochemical process.

The technology of pyrochemistry relies on the use of inert atmosphere hot cells with a very low oxygen and/or water content (1–10 ppm). The high temperature processes require the use of very sophisticated materials (Monel, Hastelloy, pyrolytic graphite, tungsten) because the molten salts and molten metals are very corrosive at high temperatures. The proposed pyrochemical infrastructure is therefore directly connected to the fuel fabrication and transmutation facility in order to limit the number of transfers and the size of the treatment plant and to restrict as much as possible transfers outside the controlled atmosphere.

On-line argon or nitrogen purification equipment is required to keep the atmospheres of the work spaces free of oxygen and moisture. Vacuum-tight levitation and transfer locks need to be installed in order to avoid diffusion of air through the plastic sleeves and joints. The chemical and electrochemical equipment inside the hot cells ought to be as vacuum tight as possible in order

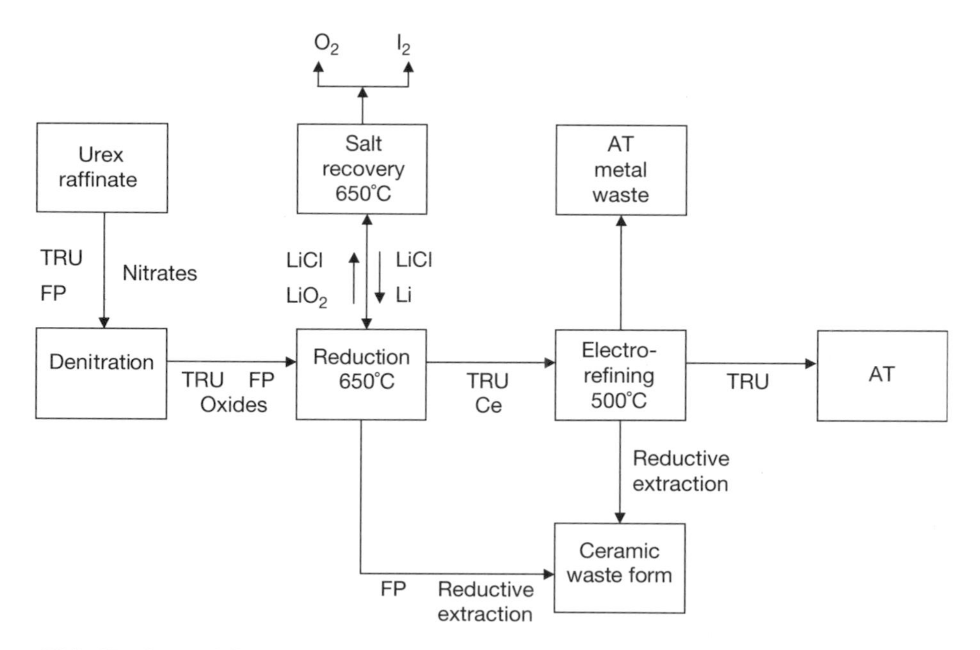


*FIG. 17. General flowsheet of the pyrochemical process. AT: accelerator transmutation.*

to limit ingress of moisture, which creates very corrosive gases in the hot cell atmosphere.

Large amounts of LiCl and chemically equivalent quantities of liquid metals (lithium) need to be introduced into the chemical reduction reactor, while its exhaust contains oxygen, $I_2$ ($^{129}I$), $^{85}Kr$ and radioactive aerosols from the salt recycling stream and other gaseous impurities. A very complex but efficient process gas purification system will need to be installed and operated in order to limit as much as possible the radioactive discharge to the surrounding atmosphere.

At the cathode of the electrochemical reactor the TRUs and some rare earth metals are separated from the fission products. Complete separation from TRUs is theoretically impossible, because some free energies of formation (praseodymium, cerium, neodymium and yttrium) overlap with those of plutonium, neptunium, americium and curium. This is not a fundamental drawback, since a certain amount of rare earths (between 1% and 10%) may be present in the TRU mixture, which is recycled in a fast neutron spectrum device. Recent studies have shown that a liquid bismuth cathode is better suited to separate the TRUs from the bulk of the rare earths. Much R&D will, however, be necessary to transform the present IFR batch process into a counter-current extraction system operating at high temperature. Technological support in upscaling may be expected from the metal refining industry, which uses similar techniques for the purification of a wide range of metals.

The waste streams discharged from the pyrochemical processes are chemically very different from the waste produced by aqueous reprocessing.

The LiCl–KCl salt waste is absorbed by a zeolite and submitted to hot isostatic compression, which operates at 850°C and 1.5 MPa pressure. The vaporization of caesium, one of the main fission products, needs to be reduced as much as possible. The glass–ceramic matrix resulting from this process is very similar in confinement characteristics to conventional borosilicate glass. Recently, the fabrication of sodalite for high active salt has been investigated, which has the advantage of low temperature operation.

The hulls discharged from the Urex process and the insoluble ‘platinum’ metals can be conditioned by melting as a homogeneous metal waste in the form of stainless steel with 15% zirconium. The homogeneity of this waste form has not yet been established. The same technique applies to a certain extent to hull waste resulting from direct pyrochemical reprocessing, although the actinide contamination of these hull fragments is not as well documented as that of hulls from aqueous reprocessing.

### 5.2.3. The European Union roadmap

In the aftermath of the US initiative on pyroprocessing, a number of European States agreed to investigate the possibilities of this technology in the context of a large scale introduction of ADS facilities to cope with sustainable long term nuclear energy development. Waste incineration is the primary goal in such a scenario and pyroprocessing a long term alternative to aqueous reprocessing and associated partitioning techniques.

The ultimate technical objective to be reached is the multiple recycling of high burnup fuel in dedicated ADS reactors and associated fuel fabrication and fuel processing facilities. Such an ambitious goal cannot be reached in a short time. Owing to the present state of development in ADS technology, MA or TRU fuel fabrication and the pyroprocessing of very high burnup fuel, an R&D period of about 30 years should be envisaged before industrial development will reach maturity.

The European Union roadmap report [5] concentrates on the development of ADS irradiation facilities, leading to the construction of an experimental ADS reactor (XADS) in 2018 and in a later phase to a dedicated MA or TRU transmutor (XADT). The fuel cycle R&D associated with this development covers the entire range of the fuel types investigated at present: MAs or TRUs in the form of oxides, nitrides and metals.

At present, only two western European research facilities, Atalante and the ITU, are capable of carrying out R&D on MAs at the level of 150 g of americium and 5 g of curium. Since there is an annual output in the European Union of 1600 kg of neptunium and 1700 kg of americium and curium, an upscaling factor of 10 000 needs to be realized.

In the European Union, the nuclear research strategy for the partitioning of MAs might become the first, and for a relatively long period the most important, research activity associated with current aqueous reprocessing. The MAs, if separated, can be stored for an extended period of time together with spent LWR MOX fuel. Immediate transmutation of the separated MAs is only of a very long term benefit, which is negligible compared with the presence of the large inventories of plutonium isotopes and MAs in stored spent LWR MOX inventories.

Separation of iodine during reprocessing is a standard procedure, but at present it is transferred to the LILW stream and discharged into the sea. Absorption of iodine on a zeolite or precipitation as an insoluble compound is an improvement. Transformation into an irradiation target is an open question that is yet to be resolved.

Technetium separated from HLLW and transformed into metal, together with insoluble technetium from the dissolver, can be transformed into a target for irradiation in LWRs. However, dedicated reactors will have to be constructed for this purpose. The problem of the LLFPs is treated in Section 6.5.

The investment for long lived radionuclide MA partitioning is relatively small compared with the whole nuclear fuel cycle, and the additional cost for the long term storage of these radionuclides is marginal compared with that for spent LWR MOX.

If the transmutation option for MAs is chosen, development of ADS systems would take several decades until they became operational at the industrial level. However, the required ADS capacity for this limited purpose represents only 5% of the total reactor capacity.

### 5.2.4. Russian development programme

More advanced stages of R&D have been reached in the Russian Federation and the Czech Republic, where pilot scale facilities have been set up and fuel has been recycled at the 5 kg level.

The most important recent (1996) technological demonstration of pyrochemical reprocessing was performed in the Russian Federation, where a sequential electrolysis process in NaCl–CsCl was demonstrated [63]. MOX fuel irradiated to ~200 GW·d/t HM in the BOR-60 reactor has been reprocessed at the RIAR centre at Dimitrovgrad using a pyroreprocessing technique, and fuel fabrication of recycled plutonium fuel has been achieved by vibrocompaction. Decontamination of noble metals from $UO_2$ deposits and co-deposition as MOX are current research issues.

In this process the spent FR MOX fuel is chlorinated by gaseous $Cl_2$ in a liquid bath of LiCl–KCl at 600–650°C. After extraction of $UO_2$ by electrolysis at the cathode, the salt bath is oxidized, leading to the precipitation of $PuO_2$. It is planned to incorporate the separation process of neptunium into this flowsheet in order to recycle TRU or plutonium–neptunium fuel into the BOR-60 reactor. This flowsheet is not applicable to americium and curium. A DF from fission products of 100 was reached in the test.

It is planned to incorporate the separation processes of MAs into the pyroreprocessing method (DOVITA) and to carry out the recycling of the TRU fuel by vibrocompaction.

The Russian approach to the development of nuclear energy is based on the continuation of the water cooled, water moderated power reactor (WWER) programme and the simultaneous development of the FR strategy, with emphasis on the use of liquid metal coolants (lead or lead–bismuth) and a variant of pyrochemical processing.

Design studies on FRs in the Russian Federation include as a first step the lifetime extension of current FRs (BR-10, BOR-60 and BN-600) and engineering design studies on BN-1800. Development of hybrid cores for BN-600 and a nitride core for BN-800 are programmed. Development of a high safety FR design (BREST-OD-300 and, later, 1200) cooled with heavy metal coolants (lead or lead–bismuth) and on-site reprocessing is the latest proposal within the Russian nuclear industry.

The lead cooled FR has a simpler design than a liquid metal fast reactor (LMFR) with sodium coolant: a single vessel configuration, decay heat removal by natural circulation and many reactivity and mechanical simplifications. Due to these new studies on the BREST-1200 reactor, a better economic competitiveness of FRs is anticipated.

### 5.2.5. Japanese research and development programme

For more than 15 years R&D has been undertaken in Japan on pyrochemical reprocessing [64] of spent oxide fuel, on HLLW and more recently on spent nitride fuel [65]. The basic principle is the same as for the Argonne National Laboratory process, but very specific additions and modifications have been brought to the IFR processes. Most of the tests have been performed with simulated fuel and with uranium and plutonium. Spent oxide fuels, especially high burnup $UO_2$ and MOX, are transformed into metals by chemical reduction followed by electrorefining for further decontamination of rare earths. HLLW is converted to chloride after calcination to separate actinides with a high recovery rate and large separation from rare earths by reductive extraction in a LiCl–KCl eutectic mixture. Thorough fundamental research has been undertaken by different universities in Japan associated with CRIEPI and JAERI [65]. A nitride fuel cycle in a double strata system with an ADS is currently being investigated at JAERI. A flow diagram of an integrated fuel cycle with an LWR and an FR with metal fuel and recovery through a pyrochemical process is shown in Fig. 18; the actinide recycling system, including the spent oxide treatment and the recovery of actinides from HLLW discharged from the Purex facility, is shown. In the FR system, spent metal fuel is treated by electrorefining and reductive extraction in order to recycle all the actinides, including neptunium, americium and curium. Injection casting is employed for the fuel slag of uranium–plutonium–zirconium–MA–rare earth. Lithium reduction is applied to transform oxide prior to the electrorefining process. Currently, electrochemical reduction, which is expected to be more effective and simple, is envisaged. Spent $UO_2$ or MOX fuel loaded in an anode basket is directly reduced by electrolysis in a chloride salt, and simultaneously some fission products, alkali, alkaline earths, etc., dissolve in the salt.

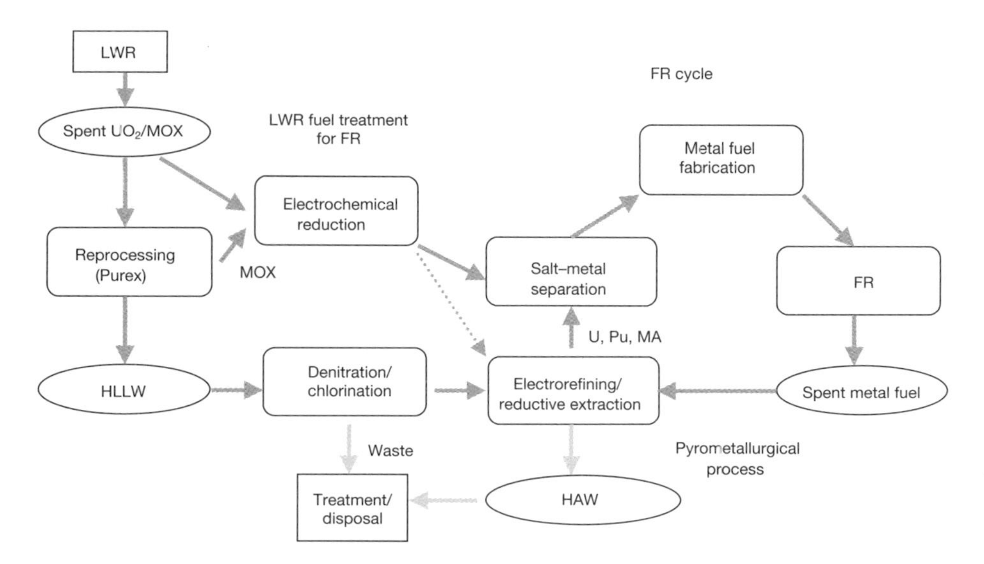


*FIG. 18. Actinide recycling system by FR integrated oxide fuel treatment and recovery of actinides in HLW.*

Provided that enough MOX is prepared, the product is added directly to plutonium and supplied to FR fuel fabrication following the separation of metal from salt. If this is not the case, the product is supplied to the electrorefining process. The HLLW has to be calcined and denitrated to convert it into oxide, which is transformed in a chloride compound by reacting the oxide with chlorine gas.

Fuel dissolution into molten salt and uranium recovery have been demonstrated at the engineering scale. Plutonium recovery using a liquid cadmium cathode has been verified by a laboratory scale installation operated in cooperation by the ITU and JAERI. Tests with MOX containing simulated fission product oxides showed lithium reduction to be technically feasible. The applicability of the effective recycling of lithium by electrolysis of $Li_2O$ remains to be addressed. For the treatment of HLLW for actinide separation, conversion to chlorides through oxides has also been established in uranium tests. Through testing it has been confirmed that more than 99% of TRU radionuclides can be recovered from HLW.

In order to verify the different process steps in real radioactive conditions, CRIEPI decided to build a specific pyrochemical laboratory infrastructure with the ITU in which all the process steps could be tested. Material selection for high temperature operation is an issue that will require extensive investigation.

### 5.2.6. Other activities

In some States several basic activities are envisaged. Experimental and theoretical work in the area of the development of pyrotechnology for ADS in the Czech Republic is directed at the fields of fluoride volatility and material research on fluoride salt. Basic studies of the thermodynamic properties of a few molten salts for electrorefining are being carried out in the Republic of Korea. Basic research on the molten salt electrorefining process for advanced fuels, such as alloy, carbide and nitride fuels, is in progress in India. A laboratory scale facility has been set up to carry out some studies using uranium.

## 5.3. SEPARATION OF LONG LIVED FISSION AND ACTIVATION PRODUCTS

A number of radiologically important fission and activation products play a potentially important role in the assessment of a geologic repository and have been considered as a P&T option. The following radionuclides have to be assessed: the fission products $^{99}Tc$, $^{129}I$, $^{135}Cs$, $^{79}Se$, $^{93}Zr$ and $^{126}Sn$ and the activation products $^{14}C$ and $^{36}Cl$.

### 5.3.1. Fission products

Caesium-137 and $^{90}Sr$ are the two main fission products that determine the radiological hazard and heat content of HLW during the initial 300 years. The radionuclide loading of the glass is determined by the concentration of these radionuclides. If they could be eliminated and conditioned in suitable matrices, the residual HLW could be disposed of much earlier in deep geological disposal facilities and the separated caesium–strontium radionuclides could be kept in engineered storage vaults.

The separation of caesium ($^{137,135,134}Cs$) from HLLW has been considered for many years. Caesium radionuclides can be effectively extracted from HLLW using several methods: adsorption on inorganic exchangers; liquid extraction with cobalt dicarbollide (CCD); and use of calixarene crown ethers. Pilot tests with CCD have been carried out in the USA, while representative industrial tests have taken place in the Russian Federation. A successful hot test with a calixarene crown ether extractant has been carried out recently at the CEA.

After separation, $^{135}Cs$ cannot be considered for transmutation because of the presence of other caesium radionuclides; separated $^{135}Cs$ will therefore be better directed towards specific conditioning into a stable crystalline matrix for disposal.

Strontium-90 separation is in itself not essential within a P&T strategy, but since it generates a large amount of heat, its removal would contribute to a decrease of the heat load in the storage formation. Hot separation tests have been performed in the Russian Federation using CCD and in India and the USA with crown ethers (DC18C6 and DtBaDC18C6).

Technetium-99 is a fission product with a half-life of 213 000 years that occurs as technetium metal and $TcO_2$ in insoluble residues and as a soluble pertechnetate ion in HLLW solution. Its generation rate is 26.6 kg/GW(e)·a, with an overall specific concentration of ~1.2 kg/t HM, depending on the burnup. In order to effectively address the long term radiotoxicity problem, both soluble (80%) and insoluble (20%) fractions ought to be combined before any nuclear action is taken towards depletion by transmutation.

The extraction of soluble $TcO_4^-$ is relatively easy. The similarity between technetium and the platinum metals in insoluble waste and the nature of the separation methods makes this partitioning operation very difficult, but separation from aqueous effluents is possible in an advanced Purex scheme. However, recovery from insoluble residues is very difficult. The present recovery yield could approach 80% at best (DF = 5).

A significant improvement of the $^{99}$Tc recovery from HLW is only possible if it is converted into a single chemical species, which is not easy to achieve. Pyrometallurgical processes are perhaps more adequate to carry out a group separation with the platinum metals.

Iodine-129 is in most of the land based repository concepts for spent fuel the first radionuclide to emerge into the biosphere, due to its very high mobility in aquifers. In spent fuel it occurs as molecular iodine, as soluble CsI, as solid $ZrI_{4-n}$ and as volatile $ZrI_4$.

Iodine as a fission element is generated in spent fuel at a level of 7.1 kg/GW(e)·a. About 80% of this inventory is present as the very long lived (16 million years) isotope $^{129}$I, and 20% as stable $^{127}$I. During reprocessing (Purex, Urex) it is removed from the dissolver solution with a yield approaching 95–98% (DF = 20–50). The radioactive concentration of this radionuclide in spent fuel is, depending on the burnup, ~$1.6 \times 10^9$ Bq/t HM and its annual limit on intake (ALI) is $2 \times 10^5$ Bq. Since its radiotoxicity is the highest among the fission products ($1.1 \times 10^{-7}$ Sv/Bq), and it is very soluble, it would be advisable to increase the separation yield from different waste streams to reduce the radiological impact. A target DF of ~1000 could be proposed as a significant improvement.

In order to improve this separation yield, more complex chemical treatments are necessary. During high temperature pyrochemical processes higher separation yields could in principle be expected. Adapted conditioning methods for separated iodine (AgI, $Pb(IO_3)_2$, Pb apatite, etc.) have been developed.

The separated fraction can either be stored on a specific (zeolite) adsorbent or discharged into the sea. Since $^{129}I$ has a half-life of 16 million years it cannot be prevented from worldwide dispersion in the geosphere or biosphere. However, in a salt dome (evaporated seawater) type of repository the dilution of eventually migrating $^{129}I$ by the mass of natural iodine ($^{127}I$) present in the body of the salt dome strongly decreases the radiological hazard.

In a worldwide dispersion scenario in the sea, its radiotoxic importance is rather limited due to its dilution by natural ($^{127}I$) iodine, as long as the LWR reprocessing capacity remains at its present level.

However, conditioning and confinement in, for example, a salt dome are the possible management options to reduce its radiological impact and final storage. This is an alternative management option that undoubtedly deserves much international attention.

Selenium-79 is a fission product (0.16 kg/GW(e)·a) with a half-life of 65 000 years that occurs in HLLW. Chemically, this radionuclide behaves as a sulphate and will be incorporated in vitrified waste. Its radioactive concentration in spent fuel is expected to be around $2 \times 10^{10}$ Bq/t HM and its ALI is $10^7$ Bq/a. Separation from liquid HLLW is not easy, owing to the very small chemical concentration in which it occurs in comparison with natural sulphur compounds.

Zirconium-93 and $^{135}Cs$ are two long lived (1.5 and 2 million years half-life, respectively) radionuclides that occur in spent fuel at relatively high concentrations (23 kg/GW(e)·a and 12.5 kg/GW(e)·a, respectively). Separation of these radionuclides from the other fission products for eventual transmutation is almost excluded, since they are accompanied by other radioisotopes that are either very radioactive ($^{137}Cs$) or present in much larger quantities (~23 kg $^{93}Zr$ among 118 kg of zirconium per GW(e)·a). In order to effectively reduce the radiotoxic potential by neutron irradiation, a series of isotopic separation processes ought to precede any target fabrication, but this route is at present considered to be an almost impossible endeavour from both the technical and economic points of view.

Tin-126, which has a half-life of 250 000 years, is partly soluble in HLLW and occurs partly in insoluble residues. Its chemical concentration amounts to 0.72 kg/GW(e)·a $^{126}Sn$ among 1.81 kg/GW(e)·a of tin in spent fuel. The radiochemical concentration of $^{126}Sn$ in HLW ranges around $3.2 \times 10^{10}$ Bq/t HM and its ALI is $3 \times 10^6$ Bq/a. The radioactive species $^{126}Sn$ is accompanied by a series of stable isotopes ($^{116,118,119,120,122,123,124}Sn$), which makes it difficult to consider for transmutation.

### 5.3.2. Activation products

Carbon-14, with a half-life of 5730 years, is problematic because it can potentially enter into the biosphere through its solubility in groundwater and play an important radiotoxicological role through its uptake into the biochemical life cycle. Its concentration in spent fuel is about $3 \times 10^{10}$ Bq/t HM and its ALI is $4 \times 10^{7}$ Bq/a, depending on the nitrogen contamination of the initial $UO_2$ fuel. Its role in long term radiotoxicity is dependent on the physico-chemical conditions occurring in the deep underground aquifer or water unsaturated geosphere. The capture cross-section in a thermal neutron spectrum is negligible.

Chlorine-36 zircaloy cladding contains some natural chlorine impurity at the level of 5–20 ppm. During irradiation, natural $^{35}Cl$ is partially transmuted into $^{36}Cl$ with a half-life of 300 000 years. This activation product arises partly in the dissolver liquid and partly remains within the washed zircaloy hulls. At 45 GW·d/t HM about $2 \times 10^{6}$ Bq is calculated to be globally present in HLW and LILW. The ALI by ingestion is $2 \times 10^{7}$ Bq/a. Owing to its chemical characteristics, this radionuclide is gradually dissolved in groundwater and could contaminate water bodies around a repository. This radionuclide cannot be considered in a recovery or transmutation scenario. The presence of natural $^{35}Cl$ in all natural waters precludes further transmutation.

### 5.3.3. Other radionuclides

Some other radionuclides discussed in this section ought to be examined in depth in order to establish their risk and potential radiotoxic role in comparison with the TRUs. Their radiotoxicity is between 1000 and 100 000 times less important than that of TRUs but their contribution to the very long term risk is predominant because migration to the biosphere may be much more rapid and generate in the very long term a non-negligible radiation dose to a human. The issues related with the very long term risk are more of an ethical than of a technological nature.

Separation of fission and activation products by pyrochemical methods has not yet been considered. Volatilization of caesium and iodine at high temperature (500°C) could in principle be envisaged.

# 6. TRANSMUTATION

## 6.1. TRANSMUTATION EFFICIENCY

The term 'transmutation' is traditionally applied to nuclear reactions that change one element into another. For the purposes of this report it is useful to broaden the concept to cover reactions changing the number of neutrons in the nucleus, such as (n,$\gamma$) and (n,2n), etc. In evaluating the transmutation efficiency of actinides, it is important to distinguish between, on the one hand, transforming a radionuclide into another actinide isotope by absorption of one or more neutrons followed by decay and, on the other hand, fissioning it into relatively short lived fission products. One can then classify transmutation reactions as direct and indirect fission. On this basis, one speaks of $^{238}$Pu(n,$\gamma$)$^{239}$Pu as indirect because a fissile isotope is generated, whereas the reactions $^{235}$U(n,f) and $^{239}$Pu(n,f) are direct fissions.

The term 'transmutation efficiency' is commonly used to describe the overall efficiency of the P&T process. It is more convenient, however, to restrict the meaning of the term to the fraction of atoms transmuted in a single irradiation step in the reactor (before reprocessing). This transmutation efficiency is determined either by experimentation or by using computer simulation. Clearly, if the transmutation efficiency is 1, all atoms are fissioned and no further reprocessing is required. In general, however, the fraction of waste atoms in a target fissioned during the irradiation period varies between 0.8 (thermal spectrum) and 0.2 (fast spectrum). In the case of the fast neutron spectrum the waste atoms have to be recycled many times in the reactor before being completely transmuted. At each reprocessing step there are partitioning losses.

The overall losses clearly depend on the number of partitioning steps required and this, in turn, depends on the transmutation efficiency. It is useful to introduce the concept of the overall efficiency of the combined P&T processes, denoted by $\varepsilon_{PT}$. This overall efficiency, following many P&T steps, is related to the individual partitioning efficiency $\varepsilon_P$ and the transmutation fraction $\varepsilon_T$ through the following [6]:

$$\text{Total losses} = 1 - \varepsilon_{PT} = R(1 - \varepsilon_P)/M$$

and

$$\varepsilon_{PT} = \frac{\varepsilon_T}{\left[1-\left(1-\varepsilon_T\right)\varepsilon_P\right]}$$

where

$M$ is the top up rate of actinides;
$R\varepsilon_P$ is the recycling rate after partitioning;
$\varepsilon_T$ is the fissioning fraction (burnup);
$\varepsilon_P$ is the partitioning efficiency.

This relation assumes that the P&T efficiencies are the same in each recycling step. As mentioned above, if the transmutation efficiency $\varepsilon_T = 1$, then the efficiency of the overall process is also 1 (i.e. $\varepsilon_{PT} = 1$). In a more realistic case, the partitioning efficiency $\varepsilon_P$ varies from 0.98 to 0.999 (e.g. for plutonium) and the optimal transmutation efficiency $\varepsilon_T$ in a fast neutron spectrum ranges between 0.15 and 0.20 (implying that 15–20% of the atoms are transmuted in a single irradiation). As a result, 80–85% of the fuel has to be recycled ($R\varepsilon_P$) five to seven times, respectively. In order to limit the overall actinide losses to the waste to 1%, the average partitioning efficiency for TRUs must attain 99.8%.

A dedicated transmutation reactor with an operating power of 1 GW(e) transforms, and fissions, during one year of continuous operation ~1 t of actinides.

At present in the European Union approximately 3.2 t of MAs are produced per year in the reactor fleet producing 100 GW(e) at a burnup of 50 GW·d/t HM [23]. This implies that at least the equivalent of three 1 GW(e) class transmuters (FRs or ADS), corresponding to 3% of the reactor fleet, are required to transform the produced MAs into fission products to stabilize the MA inventory in the reactor fleet waste. However, in a complex reactor fleet (Fig. 10) associated with a double strata strategy the MAs are usually accompanied by plutonium driver fuel, which inevitably leads to the formation of MAs. In this specific case a 6 GW(e) dedicated FR/ADS capacity needs to be installed.

In order to decrease as much as possible the required transmutation capacity (FR or ADS), the use of IMFs may be envisaged for MAs and even for the elimination of degraded plutonium (see Annex II).

## 6.2. FUEL CONCEPTS FOR TRANSMUTATION

### 6.2.1. Solid fuel

There are a wide variety of solid fuel types with a large range of chemical components, for example oxides, nitrides, carbides and metal alloys. In addition, fuel ranges from low density phases (e.g. sphere packs) to high density forms (e.g. pellets), and from homogeneous (solid solutions) to heterogeneous (cercer or cermet) materials. These can be used in critical or subcritical reactor systems for transmutation–incineration purposes. At the ITU, considerable effort has gone into the development of fabrication methods for fuels and targets. Details of the fuels and targets and fabrication methods are given in Table 2.

#### *6.2.1.1. Oxide fuels and targets*

(a) Light water reactor MOX minor actinides and fast reactor MOX minor actinides

Conventional LWR MOX fuel fabrication could produce LWR MOX 1% neptunium without major refurbishment in the MOX fabrication plant. The incorporation of $^{241,243}$Am into homogeneous LWR MOX 1% americium would be possible if additional shielding is put around the gloveboxes or in (semi-) automatic production units. However, this option can only be taken if a separation of americium and curium has been realized. The presence of the slightest trace of $^{244}$Cm induces an important neutron and gamma background [66].

The FR MOX MA fuel fabrication can rely on the experience gained during FR MOX fuel fabrication. However, due to the type of fuel fabrication (powder mixing, pelletizing, sintering) and the use of existing facilities, curium is also excluded from handling in a conventional FR MOX fuel fabrication plant. From the reactor physics point of view, the MA content of the fuel can be increased to 2.5% neptunium or americium without a major impact on reactor safety [67]. Homogenous recycling of neptunium is the preferred option, while the preference for americium is for heterogeneous recycling, which would allow fabrication of americium pins separately from the bulk of FR MOX fuel pins.

The presence of significant concentrations of americium requires novel production methods, for example sol gel, infiltration and vibropac techniques. These methods are under active development in the European Union and the Russian Federation [63, 68].

(b) Fertile free fuel or targets

The advantage of this IMF utilization in LWRs or FRs lies in a better plutonium consumption as compared with MOX. R&D programmes on IMF have led to the selection of YSZ doped with erbia and plutonia at the PSI [41] and the deployment of experimental projects including material preparation

TABLE 2. PROGRAMME OF TRANSMUTATION–INCINERATION, SHOWING FUELS AND TARGETS FABRICATED, AT THE INSTITUTE FOR TRANSURANIUM ELEMENTS

| Programme | Reactor | Fuel/target | Method | Status |
|---|---|---|---|---|
| FACT | FR2 (1981) | $(U_{0.5}Am_{0.5})O_2$ | Sol gel | PIE[a] complete |
| MTE2 | KNK II (1984–1985) | $NpO_2$; $(U_{0.5}Am_{0.5})O_2$ | Sol gel | PIE complete |
| | | $(U_{0.73}Pu_{0.25}Np_{0.02})O_2$ | Sol gel | PIE complete |
| | | $(U_{0.73}Pu_{0.25}Am_{0.02})O_2$ | Sol gel | PIE complete |
| SUPERFACT1 | Phenix (1986–1988) | $(U_{0.74}Pu_{0.24}Np_{0.02})O_2$ | Sol gel | PIE complete |
| | | $(U_{0.74}Pu_{0.24}Am_{0.02})O_2$ | Sol gel | PIE complete |
| | | $(U_{0.55}Np_{0.45})O_2$ | Sol gel | PIE complete |
| | | $(U_{0.6}Am_{0.2}Np_{0.2})O_2$ | Sol gel | PIE complete |
| POMPEI | HFR[b] (1993–1994) | Tc metal | Casting | PIE in progress |
| | | Tc–50% Ru metal | Casting | PIE in progress |
| | | Tc–80% Ru metal | Casting | PIE in progress |
| TRABANT1 | HFR (1995–1996) | $(U_{0.55}Pu_{0.40}Np_{0.05})O_2$ | Sol gel | PIE in progress |
| | | $(Pu_{0.47}Ce_{0.53})O_{2-x}$ | Sol gel | PIE in progress |
| EFTTRA-T1 | HFR (1994–1995) | Tc metal (three pins) | Casting | PIE complete |
| EFTTRA-T4 | HFR (1996–1997) | $MgAl_2O_4$–12% Am | INRAM (pellets) | PIE complete |
| EFTTRA-T4bis | HFR (1997–1999) | $MgAl_2O_4$–12% Am | INRAM (pellets) | PIE in progress |
| TRABANT2 | HFR | $(U_{0.55}Pu_{0.45})O_2$ | Sol gel | Irradiation to be started |
| | | $(U_{0.6}Pu_{0.4})O_2$ (two pins) | Mechanical mixing | Irradiation to be started |

TABLE 2. (cont.) PROGRAMME OF TRANSMUTATION–INCINERATION, SHOWING FUELS AND TARGETS FABRICATED, AT THE INSTITUTE FOR TRANSURANIUM ELEMENTS

| Programme | Reactor | Fuel/target | Method | Status |
|---|---|---|---|---|
| ANTICORP | Phenix | Tc metal | Casting | Irradiation to be started |
| METAPHIX | Phenix | U, Pu, Zr | Casting | Irradiation to be started |
| | | U, Pu, Zr, MA 2%, rare earth 2% | Casting | Irradiation to be started |
| | | U, Pu, Zr, MA 5%, rare earth 5% | Casting | Irradiation to be started |
| | | U, Pu, Zr, MA 5% | Casting | Irradiation to be started |

[a] PIE: Post-irradiation examination.
[b] HFR: High Flux Reactor.

[69], irradiation tests on inert matrices utilizing accelerators and tests on the IMF within research reactors. Several studies summarize the results obtained so far for zirconia implantation with iodine, xenon and caesium as representative fission products, including reactor tests performed in the HFR at Petten [70] and in the HBR at Halden [71].

Fabrication of americium target pins based on the impregnation technique (INRAM) has been undertaken at the ITU. Porous $ZrO_2$ has been loaded with 10–20% $AmO_2$ for irradiation in the HFR. Higher americium enrichments, up to 40%, are being investigated [72]. The main issue to be solved is the homogeneous enrichment of the porous support.

#### 6.2.1.2. *Non-oxide fuels and targets*

Non-oxide materials are fuel or target components that can be used as fertile or fertile free material in fast systems. The advantages of these materials are that they have a higher density and better thermal conductivity than oxides. The higher density results in a decrease in the moderation and consequently a harder neutron spectrum. The higher thermal conductivity allows lower peak temperatures in the reactor with improved safety characteristics.

(a) Fast reactor metal and nitride fuels

A uranium–15% plutonium–10% zirconium alloy with a melting point of 1180°C is the preferred fuel form for the IFR reactor. It is manufactured by casting the electrochemically obtained uranium–15% plutonium–10% zirconium alloy at 1300°C in stainless steel fuel pins. In principle, it might be possible to use the same technology to produce FR metal plutonium–MA fuel, but reactor physics criteria limit the MA concentration to 5% in sodium cooled reactors. Electrorefining is under investigation for the reprocessing of such fuels.

Nitride fuel has been investigated because of its high thermal conductivity and its inertness towards liquid sodium and lead. The $^{15}N$ isotope has also been considered for use in nitride fuels. Owing to its low neutron absorption cross-section, it does not lead (in contrast to $^{14}N$) to the production of environmentally hazardous $^{14}C$ through (n,p) reactions. Nitrogen-15 is, however, very expensive to produce. The technology to produce nitride fuel is still at the laboratory stage [73]. Nitrides of TRUs can be anodically dissolved in a molten LiCl–KCl bath at 500°C. Electrorefining has been carried out in small laboratory conditions. The main issue is the use and recovery of the (expensive) enriched $^{15}N$ during the fuel processing and recycling operations. From a waste management point of view it has to be stressed that, if produced with natural nitrogen, the generation of $^{14}C$ in spent nitride fuel by the (n,p) reaction would have important environmental consequences during recycling. Pyrochemical processing seems to be the preferred recycling technology.

(b) Fertile free fuels and targets

The fabrication of (Pu, Zr)N fuel utilizing a chemical precipitation technique for oxyhydroxide followed by nitration has been undertaken at the PSI in the framework of the CONFIRM project. Pellets of zirconium nitride loaded with 10–20% plutonium nitride have been produced for irradiation in the Studsvik reactor.

### 6.2.2. Liquid fuel

Two basic types of liquid reactor fuel were investigated and tested in the early stages of nuclear power development: water solutions of uranium salts and molten uranium salts (fluorides). Molten salt fuels look attractive for transmutation, owing to the possibility of on-line pyroprocessing and continuous correction of fuel composition. Two small pilot molten salt reactors with on-line radiochemistry were constructed in the USA in the 1960s [74] and operated successfully for a few years. The transmutation potential of molten salt systems (both critical reactors and ADSs) are under investigation in the USA [75], the European Union [76] and the Russian Federation [77].

## 6.3. TRANSMUTATION POTENTIAL OF VARIOUS REACTOR TYPES

Depending on the fuel cycle and partitioning scenario, different types of reactor must be considered for transmutation purposes. Starting with the most commonly used LWRs, the most promising concepts are briefly described here.

The first stage of closing the nuclear fuel cycle is concerned mainly with plutonium transmutation. It is a well established technology based on the use of MOX fuel in European LWRs. If plutonium is involved in multiple recycling, its use in every successive cycle will be more and more similar to MA transmutation, due to the accumulation of plutonium isotopes that resemble MAs as far as the nuclear properties are concerned.

There are two main options for using the reactor fleet for transmutation. One is to involve many power reactors in the process and the other is to concentrate transmutation in a limited number of dedicated facilities with special features (the double strata fuel cycle).

### 6.3.1. Light water reactors

With the use of uranium fuel in LWRs, neutron capture reactions in $^{238}U$ give rise to plutonium. It is this plutonium that is the essential component of nuclear waste and the main source of radiotoxicity. To mitigate this problem, plutonium transmutation in MOX fuel has for several decades been realized in the European Union on the industrial scale, due to the existence of the two large reprocessing plants at Cap de la Hague in France and Sellafield in the UK, which together handle 2400 t HM of LWR $UO_2$ and 1100 t HM of gas cooled reactor uranium. This process capacity is sufficient at present to cover the needs of the European Union and some overseas customers. The use of separated plutonium was initially started to feed an FR programme, but has been delayed or stopped due to technical and economic constraints. The available reprocessing capacity has a potential throughput of about 25 t of plutonium per year.

At present, the use of separated plutonium is almost exclusively connected with the production of LWR MOX fuel, which is used as a substitute for LWR $UO_2$ fuel. However, the delay in the construction of LWR MOX facilities has led after several decades to an increased fraction of stored separated plutonium. At present, the overall LWR MOX fuel fabricating capacity in the European Union is about 300 t HM (~25 t of plutonium per year).

The LWR MOX fuel output with plutonium enrichment varies from ~5% to ~8%, depending on the fissile isotopic content and on the delay between reprocessing and fuel fabrication. It is used in thermal reactors in a proportion of 20–33% of a reactor core in several European States. The balance between the consumption and production of plutonium in the reactor core depends on the burnup of the fuel. In the case of a reactor core loaded with 1/3 LWR MOX fuel and 2/3 LWR $UO_2$ and irradiated until 50 GW·d/t HM, the balance between the plutonium consumption in the MOX fuel and production in the $UO_2$ fuel is slightly negative, but if the production of MAs in both parts of the core is being considered there is an increase in the overall TRU content. The inventory reduction of plutonium in irradiated LWR MOX amounts to 33%, but almost 10% is transformed into MAs. The net TRU inventory reduction is in this case ~24%. When totalling the generation of TRUs in such a recycling process the overall radiotoxicity increase is small but not negligible (~9%). Multiple recycling of plutonium in LWR MOX is not efficient in terms of radiotoxicity reduction [78].

The radiotoxic inventory of spent LWR MOX fuel is about eight times higher than that of spent LWR $UO_2$. Conventional reprocessing will remove uranium and plutonium, which accounts for about 30% of the total alpha activity, and the residual 70%, which is made up of neptunium, americium and curium, enters the HLLW. However, the curium and americium isotopes constitute the overwhelming majority of this alpha activity. In the case of a core loading with 1/8 LWR plutonium IMF and 7/8 LWR $UO_2$, the plutonium production balance is reached [79].

Another option is to store the spent LWR MOX fuel, for example over 50 or more years, and to let the $^{244}Cm$ decay (18 years half-life) to $^{240}Pu$ before carrying out the reprocessing. The chemical extraction processes are much easier to perform after the extended 'cooling' period, as the alpha decay heat is reduced by a factor of seven or more, depending on the isotopic composition. However, this option involves the continued availability of the reprocessing facilities.

#### *6.3.1.1. Minor actinide recycling in light water reactors*

The MAs constitute the main source of radiotoxicity in vitrified HLW. Their confinement from HLLW produced during advanced reprocessing and conditioning in an insoluble matrix is a first, but not sufficient, step. In the risk reduction of long lived waste, transmutation of long lived MAs, with formation of shorter lived radionuclides, and 'incineration', with formation of much less toxic fission products, could be a second step that would be crucial to reducing the long term radiotoxicity.

Homogeneous recycling of MAs in LWR MOX fuel is a practice that does not keep the surplus MA source term separated from the driver fuel and would further dilute the MAs, especially $^{237}Np$, in a larger volume of spent LWR MOX fuel, which would be very difficult to reprocess later, because of $^{238}Pu$ accumulation. Heterogeneous recycling of MAs in the form of irradiation targets or individual fuel pins in an LWR $UO_2$ or LWR MOX core is an alternative that is much more attractive from a fuel cycle point of view, since target fabrication and chemical processing are independent of the industrial large scale operations on spent fuel.

Reduction of the radiotoxic inventory of nuclear material can more efficiently be achieved from the neutron economics point of view by performing transmutation in a thermalized fast spectrum device (an FR or ADS).

A comparative study of transmutation in LWRs compared with FRs and ADSs has been undertaken at SCK•CEN [80] and the ITU [9] in order to assess the spent target composition as a function of neutron fluence. The SCK•CEN study concentrated its scope on the comparatively short term effects of MA irradiations in a PWR, a HFR and an ADS, while the ITU emphasized long term 'one shot' irradiations in a PWR and an FR.

The main conclusions of this study shed some light on the successive stages of the overall transmutation process. While during the early phase of the irradiation stages the disappearance of the MAs in thermal systems (LWR MOX, MTR-BR2) mainly consists of transmutation to higher TRU isotopes, it gradually evolves with increasing fluence towards incineration as the role of the secondary fissions increases.

The difficulty of handling highly irradiated MA targets in a multiple recycling operation led to the consideration of once through irradiation as a possible option. The European Union produces 1.6 t of neptunium and about 1.8 t of americium and curium per year. For reactivity reasons, only 1% neptunium or americium and curium may be loaded in a large LWR core. The addition of 1% MAs in an LWR core would call for an equivalent increase of $^{235}U$ enrichment.

#### *6.3.1.2. Irradiation of neptunium*

Homogeneous irradiation of 1% neptunium in standard PWR $UO_2$ fuel to a burnup of 47 GW·d/t HM leads to a depletion of 45–50% (i.e. a reduction factor of only two is obtained). Further reduction of the neptunium inventory would require repeated and selective reprocessing of this neptunium doped fuel, which is a complex and very expensive process.

Irradiation of $^{237}$Np in LWRs leads during the first three years to the production of $^{238}$Pu, with some minor $^{239}$Pu and fission product impurities (see Fig. 19 and Table 3). With increasing fluence during long term irradiations the $^{238}$Pu is transformed into $^{239}$Pu and mainly fissioned. Theoretically, an incineration rate of 90% is reached after 16 years of irradiation. The residual actinide content of about 10% is made up of $^{244}$Cm and some traces of $^{239,240,242}$Pu. If such a target decays, it is gradually transformed into 90% fission products and 10% plutonium isotopes ($^{240}$Pu, progeny of $^{244}$Cm and the other long lived plutonium isotopes). This transmutation–incineration process reduces the actinide content by a factor of ten. However, at a total fluence of $1.5 \times 10^{23}$ n/cm$^2$ the capsule cladding may possibly fail and intermediate re-encapsulation may be required.

TABLE 3. ATOM PER CENT CONCENTRATION OF A 20% NEPTUNIUM-237 TARGET IRRADIATED IN A 25% MOX FUELLED PWR, BR2 AND MYRRHA (ADS) DURING 200, 400 AND 800 EFPD [80], FOLLOWED BY FIVE YEARS OF COOLING

| Radionuclide | PWR ($6.9 \times 10^{14}$ n·cm$^{-2}$·s$^{-1}$) | | | BR2 ($1.2 \times 10^{15}$ n·cm$^{-2}$·s$^{-1}$) | | ADS ($3.3 \times 10^{15}$ n·cm$^{-2}$·s$^{-1}$) | |
|---|---|---|---|---|---|---|---|
| | 200 EFPD | 400 EFPD | 800 EFPD | 400 EFPD | 800 EFPD | 400 EFPD | 800 EFPD |
| Fission | 1.6 | 5.3 | 17.9 | 49.4 | 81.4 | 5.9 | 12.2 |
| Np-237 | 80 | 64 | 41 | 22.1 | 4.9 | 84 | 70.5 |
| Pu-238 | 16 | 24.9 | 30.4 | 16 | 4.4 | 9.4 | 15.5 |
| Pu-239 | 1.4 | 3.3 | 5.2 | 5 | 1.5 | 0.3 | 0.9 |
| Pu-240 | 0.2 | 0.5 | 1 | 1.6 | 0.8 | 0 | 0 |
| Pu-241 | 0.1 | 0.4 | 1.3 | 1.5 | 0.5 | 0 | 0 |
| Pu-242 | 0 | 0.1 | 0.6 | 2 | 1.8 | 0 | 0 |
| Am-241 | 0 | 0.1 | 0.4 | 0.4 | 0.1 | 0 | 0 |
| Am-243 | 0 | 0 | 0.2 | 0.8 | 1.5 | 0 | 0 |
| Cm-244 | 0 | 0 | 0.1 | 0.3 | 1.5 | 0 | 0 |
| Total | 100 | 99.7 | 99.6 | 99.8 | 98.6 | 100 | 99.9 |

EFPD: effective full power days.

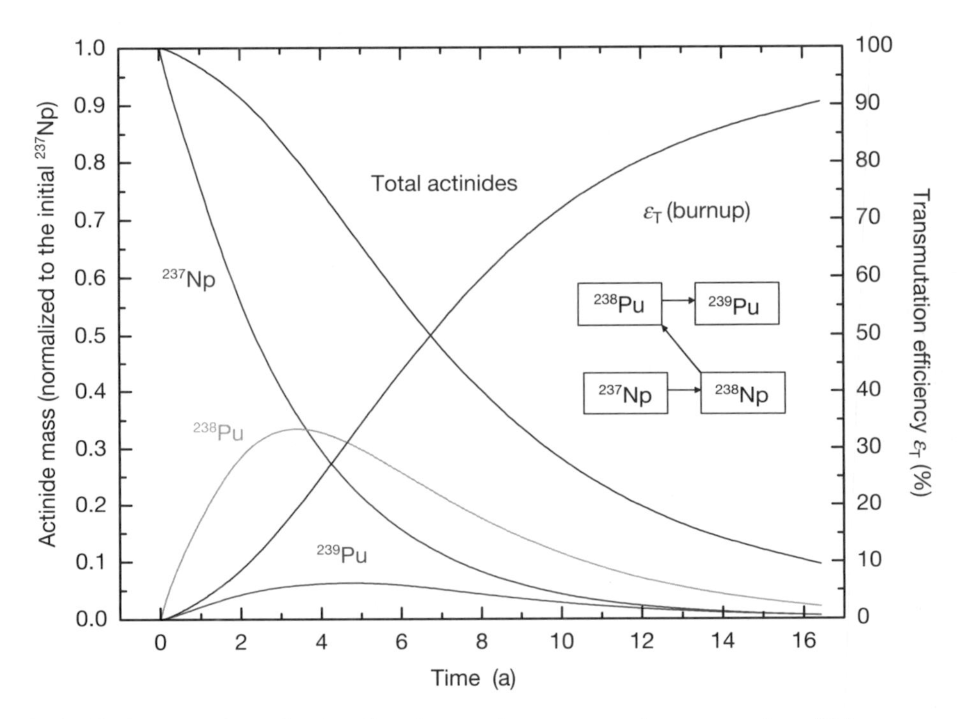


*FIG. 19. Composition of an irradiated neptunium target and transmutation efficiency as a function of fluence in a PWR [9].*

Nearly quantitative conversion (~94%) of $^{237}$Np into $^{238}$Pu is an alternative that can be realized within three cycles of approximately three years each and intermediate reprocessing at each discharge. This option reduces the (very) long term radiological impact of a repository but increases the medium term radiotoxicity of the target by a factor of $2.4 \times 10^4$ for a period of several hundred years. With a $^{238}$Pu half-life of 87 years, the target has to be kept in complete confinement during approximately ten half-lives. From a technological point of view, the $^{237}$Np–$^{238}$Pu conversion has important consequences for heat dissipation in a repository. The decay heat of $^{238}$Pu amounts to 0.57 MW/t $^{238}$Pu.

#### *6.3.1.3. Irradiation of americium*

Irradiation of americium is a much more complex nuclear reaction scheme, as two different isotopes ($^{241}$Am and $^{243}$Am) are transformed into $^{242}$Cm and $^{244}$Cm; however, owing to the short half-life of $^{242}$Cm (162 days), this radionuclide decays during its stay in the reactor to $^{238}$Pu and is transmuted in its turn into $^{239}Pu$. The discharged target after three years of irradiation, as shown in Table 4, still contains 27% americium, 60% curium and plutonium and 13% fission products.

Fabrication of special 1% americium enriched LWR $UO_2$ fuel assemblies, followed by their irradiation in a conventional PWR, could be a preliminary step in the gradual decrease of this very radiotoxic element. However, from the radiotoxicity and heat generation points of view this practice can only be an intermediary step in the americium inventory reduction, which has to be completed by irradiations in FRs or ADSs. Moreover, from the economic point of view, the inclusion of 1% americium in LWR $UO_2$ fuel requires an additional enrichment of 1% $^{235}U$.

TABLE 4. ATOM PER CENT CONCENTRATION OF A 20% AMERICIUM TARGET (78% $^{241}Am$; 22% $^{243}Am$) IRRADIATED IN A 25% MOX FUELLED PWR, BR2 AND MYRRHA (ADS) DURING 200, 400 AND 800 EFPD [80], FOLLOWED BY FIVE YEARS OF COOLING

| Radionuclide | PWR ($6.54 \times 10^{14}$ n·cm$^{-2}$·s$^{-1}$) | | | BR2 ($1.05 \times 10^{15}$ n·cm$^{-2}$·s$^{-1}$) | | ADS ($3.37 \times 10^{15}$ n·cm$^{-2}$·s$^{-1}$) | |
|---|---|---|---|---|---|---|---|
| | 200 EFPD | 400 EFPD | 800 EFPD | 400 EFPD | 800 EFPD | 400 EFPD | 800 EFPD |
| Fission | 2.6 | 5.9 | 14.3 | 24.5 | 55.7 | 4.8 | 10 |
| Np-237 | 0.5 | 0.4 | 0.2 | 0.1 | 0 | 0.6 | 0.6 |
| Pu-238 | 15.3 | 24.4 | 30.5 | 30.6 | 11.4 | 6.4 | 10.7 |
| Pu-239 | 0.2 | 1.2 | 3.8 | 4.7 | 3.3 | 0.1 | 0.4 |
| Pu-240 | 0.7 | 1.3 | 2.1 | 2.9 | 2.9 | 0.4 | 0.8 |
| Pu-241 | 0 | 0.1 | 0.5 | 0.8 | 0.9 | 0 | 0 |
| Pu-242 | 3 | 4.3 | 4.7 | 5.4 | 3.1 | 1.4 | 2.5 |
| Am-241 | 78.08 | 54.4 | 38.2 | 5.3 | 0.6 | 64.5 | 53.8 |
| Am-242m | 0.6 | 0.5 | 0.2 | 0 | 0 | 0.8 | 1.1 |
| Am-243 | 18.3 | 15.9 | 12.7 | 12.6 | 7.5 | 18.9 | 16.3 |
| Cm-243 | 0.2 | 0.4 | 0.5 | 0.7 | 0.2 | 0 | 0.1 |
| Cm-244 | 3.2 | 5.4 | 7.8 | 9.6 | 10.6 | 1.7 | 2.9 |
| Cm-245 | 0.3 | 0.8 | 1.8 | 0.9 | 1.1 | 0.1 | 0.2 |
| Cm-246 | 0 | 0.1 | 0.3 | 0.4 | 1.2 | 0 | 0 |
| Total | 99.9 | 99.9 | 99.8 | 99.7 | 99 | 100 | 99.9 |

Long term irradiation of $^{241,243}$Am proceeds in a similar fashion to neptunium irradiation, as shown in Fig. 20. A maximum incineration rate of ~90% is achieved after 16 years of irradiation, leading to a total fluence of $1.5 \times 10^{23}$ n/cm$^2$. A residual actinide content of $^{238}$Pu and $^{244}$Cm constitutes the medium term radiotoxicity (250–870 years). Since all the actinides of the mixture in the irradiated capsule have nearly the same radiotoxicity, such a transmution–incineration process reduces the radiotoxicity by a factor of ten at discharge from the reactor. The long term reduction of the radiotoxicity depends on the final $^{240}$Pu and $^{234}$U content, which is generated by $^{244}$Cm and $^{238}$Pu decay.

*6.3.1.4. Conclusions on thermal neutron irradiation of minor actinides*

Conventional nuclear power plants cannot be used to decrease significantly the neptunium and americium radiotoxic inventories resulting from LWR $UO_2$ reprocessing, since neutron capture reactions dominate the nuclear transmutation processes during the relatively short irradiation cycles. The

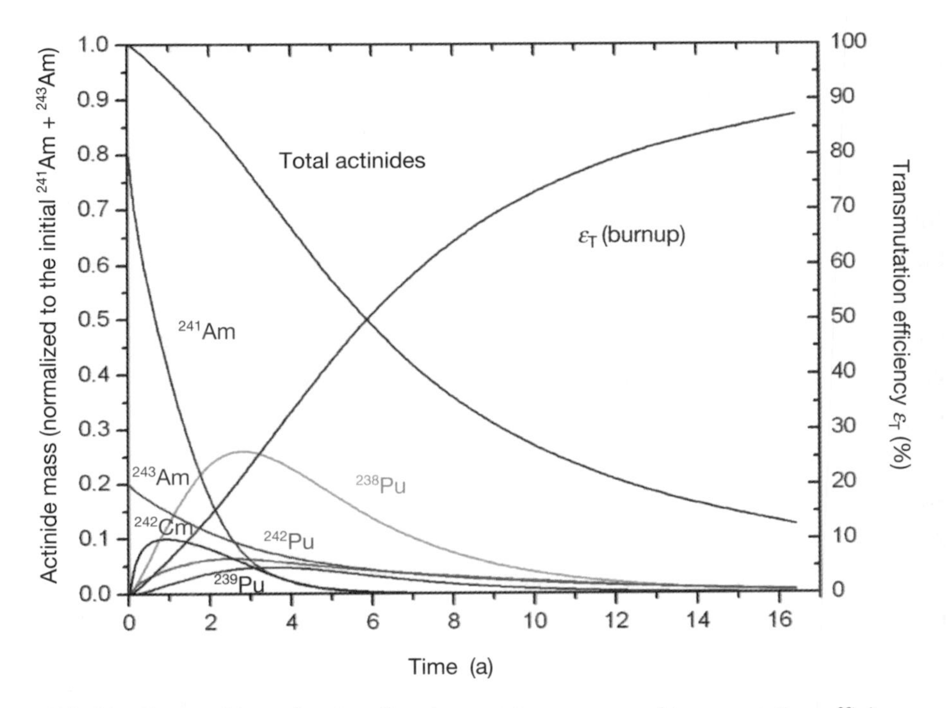


*FIG. 20. Composition of an irradiated americium target and transmutation efficiency as a function of fluence in a PWR [9].*

presence of highly active alpha emitters (especially $^{244}$Cm) in irradiated targets make multiple recycling nearly impossible.

Dedicated LWR transmutation reactors could serve as intermediary storage plants for separated actinides if an additional enrichment of 1% $^{235}$U has been provided. The low MA transmutation capability of thermal reactors has been demonstrated by different R&D groups in the European Union, which have shown that significant results could be obtained by following once through irradiation with very long irradiations (16 years) and with higher $^{235}$U enrichment.

### 6.3.2. Fast reactors

In principle, a much more efficient elimination of the actinide inventory could be accomplished by fast neutron irradiation in either a critical FR or subcritical ADS. Since all the actinides (even and uneven masses) are to a certain extent fissionable in a fast neutron spectrum, it was anticipated that irradiation in the core of an FR or ADS would result in clean incineration of the actinides. However, the fast cross-sections of most actinides are so much lower than the thermal ones that secondary neutron capture in a thermal spectrum is a much more efficient transmutation process than direct fission by a fast fission neutron.

However, the neutron economy of an FR is much more favourable than that in a thermal spectrum, since each fast neutron produces directly another fission, while in a thermal spectrum two or more neutrons are necessary to achieve fission. In order to investigate this phenomenon SCK•CEN [80] and the ITU [9] investigated the incineration capacity of a generic FR and a typical ADS (MYRRHA).

The real net radiotoxicity reduction results from fission and is consequently proportional to the burnup of the fuel or target. High burnups are the prerequisite condition to realize high radiotoxicity reduction levels. The maximum burnup obtainable in a fast neutron spectrum device (FR or ADS) is in principle limited not by the fissile material content, since all TRUs are to a certain extent fissionable, but by the resistance to fast neutron radiation damage of the fuel/target cladding material expressed in displacements per atom. If the fast neutron dose exceeds 200 dpa the stability of the cladding material is at stake and the highly irradiated fuel or target (150–250 GW·d/t HM) will have to be withdrawn from the reactor and the residual 75–85% of the initial target recycled. Multireprocessing operations with highly active targets are very difficult and should be avoided.

An alternative, which has received much attention, is irradiation in a moderated zone of an FR or ADS. By combining fast neutron economy (one fission per neutron) with the high thermal cross-section of a moderated neutron target, very high depletion rates can be achieved.

#### *6.3.2.1. Transmutation–incineration of neptunium*

Homogeneous recycling of neptunium in a typical fast reactor has little impact on the neutronics of the core as long as the neptunium content remains below 2.5%. Such a reactor could be used as a storage reactor, since approximately 1 t of neptunium could be accumulated in the core. As a result, a 1.6 GW(e) reactor would be capable of absorbing all the neptunium resulting from the reprocessing of all LWR $UO_2$ spent fuel produced annually by a 100 GW(e) grid. However, this option will not significantly deplete the neptunium content, since the fuel has to follow the reactor cycles and has to be periodically discharged from the reactor.

Transmutation of neptunium targets in FRs at a burnup of 120 GW·d/t HM yields a depletion of 60% but a fission rate of only 27%. In the burnup bracket of 150–250 GW·d/t HM the depletion yield will further increase but the final composition will be determined by the fission to capture ratio of $^{237}Np$. Long term uninterrupted irradiations, as illustrated in Fig. 21, do not yield significant depletion factors. Unless multiple recycling of the targets (three to four cycles with intermediate processing) is carried out, no significant depletion can be achieved. Reprocessing of small irradiated targets is possible in laboratory type dedicated facilities, in which either fast contactors are used for aqueous extraction or, possibly in the future, pyrochemical separation methods are employed. By multiple recycling, the neptunium inventory of the targets could be decreased by a factor of ten after five recycles. This is the price to be paid to decrease a very long term hazard without any benefit in the short and medium term.

#### *6.3.2.2. Transmutation of americium*

Since $^{241}Am$ is the precursor of $^{237}Np$, and constitutes the highest radiotoxic potential after the plutonium isotopes, its transmutation is also defendable from the radiotoxic point of view. Great progress has been achieved in the separation of americium and curium from HLLW, and it may be anticipated that concentrates of $^{241,242,243}Am$ and $^{242,243,244}Cm$ (with 10–50% rare earths) might be produced in an extension of the reprocessing plant. For obvious reasons it is preferable to keep the separated americium and curium fraction in targets (mixed, for example, with $Al_2O_3$) to be irradiated at a later stage. However, fabrication of americium and curium targets is the main

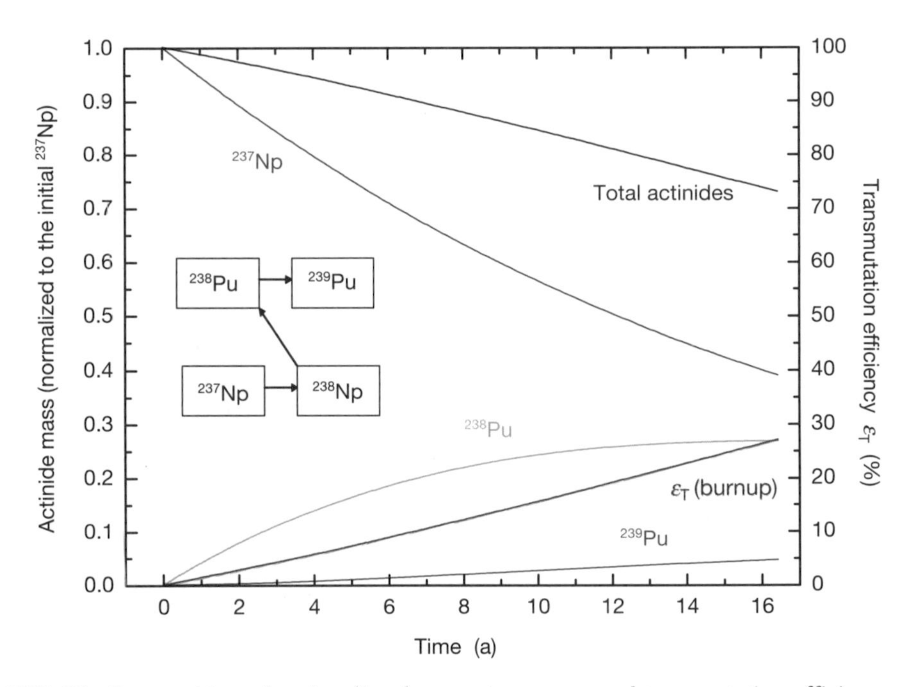


*FIG. 21. Composition of an irradiated neptunium target and transmutation efficiency as a function of fluence in an FR [9].*

technological problem to be overcome before any significant irradiation campaign can be envisaged.

Transmutation–incineration of the americium fraction in, for example, a $ZrH_2$ or $CaH_2$ moderated peripheral core position is the best way to reduce the americium inventory in a once through irradiation up to the limit of cladding resistance. In long term irradiations (15–20 years), 90–98% of the initial americium inventory can be incinerated (i.e. nearly quantitatively transformed into fission products and some percentages of $^{238}$Pu, $^{244,245}$Cm, etc.). In the specific study carried out by the CEA and Electricité de France within the European Commission research framework programme on P&T, the americium targets were considered to be irradiated in a CAPRA type reactor. At 98% americium depletion, the residual target contains the following radionuclides: 80% fission products, 12.1% plutonium, 2.16% americium and 5.2% curium [81].

The conclusion to be reached from these computations is that the (once through) irradiation of americium targets provides a solution to decreasing the americium–(neptunium) inventory by a factor of ~50, but still contains 2% plutonium isotopes. The actinide reduction factor is again between eight and ten. The radiotoxicity of the discharged target decreases by a factor of only 2.5 after 100 years of cooling. The main issues to be resolved are the target/fuel fabrication and the FR technology. Figure 22 shows the expected post-irradiation composition in a small dedicated prototype FR.

#### *6.3.2.3. Multiple recycling of plutonium and minor actinides in critical fast reactors*

A combined irradiation of plutonium and MAs is an approach that has been investigated in several reactor technology projects: the metal fuelled Integral Fast Reactor at Argonne National Laboratory [18], the MA burner reactors at JAERI [29], the CAPRA reactor project at CEA Cadarache [82] and the MA MOX fuelled advanced liquid metal reactor [83].

Each of these approaches displays a number of advantages and disadvantages, which can be summarized as follows:

(a) The fuel types and configurations aim at the generation of a neutron spectrum that is as hard as possible;
(b) Metal or nitride fuels are slightly better than oxide fuel, but are difficult to produce;
(c) The plutonium concentration increases from 33% to 45% to compensate for the anti-reactivity of the MAs;
(d) The void reactivity coefficient is positive with plutonium as the fuel and sodium as the coolant;
(e) Substitution of plutonium by $^{235}$U and sodium by lead greatly improves the reactor safety parameters.

This set of improved irradiation conditions has resulted in an expected increase of the incineration yield of TRUs but has shown that, to reach a significant (factor >10) TRU depletion, multiple recycling cannot be avoided.

Multiple recycling of plutonium and MAs from high burnup FR fuel is still at the conceptual stage and important progress has to be made before this option can be implemented. Computations have shown that an isotopic equilibrium state is reached from the fifth recycle. Aqueous reprocessing is probably inefficient for this purpose because long cooling times (10–12 years) between each recycling step are necessary to reduce the alpha radiation damage due to $^{244}$Cm on the aqueous extractants. Pyro(electro)chemical reprocessing is in the long term the only viable option for multiple recycling with short cooling times, but this technology is still in its infancy and encounters severe material problems.

The major reactor problem encountered in the design of dedicated FRs for MA burning is the positive void reactivity coefficient when the reactor is sodium cooled; the MA content must therefore be kept below 2.5%. As a consequence, a large electronuclear FR capacity (25–50% of the fleet) would be necessary to first stabilize the plutonium and MA inventory and then to decrease the overall actinide level.

## 6.4. ACCELERATOR DRIVEN SYSTEMS

The idea of using accelerators to produce fissionable material was put forward by G.T. Seaborg in 1941. He produced the first human-made plutonium using an accelerator. After that, considerable work on this type of system was performed in Europe, Japan and the USA. In the 1990s C. Bowman's Los Alamos National Laboratory group [75] and C. Rubbia's group at the European Organization for Nuclear Research (CERN) [84] designed a transmutation facility using thermal neutrons (Bowman's group) and fast neutrons (Rubbia's group), for burning both the actinides and long life fission products from spent LWR fuel. The facility, called an ADS, combines high intensity proton accelerators with spallation targets and a subcritical core with or without a blanket.

Various concepts of ADS have been proposed, with different goals and approaches. Relevant R&D programmes are being pursued at, for example, CEA, JAERI, Los Alamos National Laboratory and CERN. In recent years all the system concepts proposed by these groups have converged on a fast neutron spectrum and a gas or liquid metal coolant; most research at present concentrates on the use of lead or lead–bismuth. The advantages of lead alloy over sodium as a coolant are related to the following basic material characteristics: chemical inertness with air and water; higher atomic number; low vapour pressure at operating temperatures; and high boiling temperature.

The proton accelerator will be either a linear accelerator (linac) or a circular accelerator (cyclotron). A high intensity continuous wave proton beam with an energy of around 1 GeV and a current of several tens milliamperes is injected into a target of heavy metal. This results in a spallation reaction that emits neutrons, which enter the subcritical core to induce further neutron cascades and nuclear reactions. The subcritical core can, in principle, be operated with either a thermal or fast neutron spectrum.

ADSs have unique features for burning MAs and LLFPs, preferably in the double strata option. They operate in a subcritical mode and can more easily address the safety issues associated with criticality than critical systems. They also offer substantial flexibility in overall operation. ADSs can provide more excess neutrons than critical reactors. These excess neutrons may be utilized for transmutation, conversion and breeding purposes. These features may be exploitable for a safe and efficient means of transmuting nuclear waste. Both homogenous and heterogeneous fuel recycling is possible.

### 6.4.1. Spallation target

The spallation neutron energy spectrum is dominated by evaporation neutrons (about 90%) with energies of a few mega-electronvolts from de-excitation of reaction residues and has a tail of high energy neutrons up to the full energy of protons from pre-compound reactions within the target nuclei. Typically, a few tens of neutrons are produced by each incident proton. A proton current (5–10 mA) with 1 GeV proton energy will give rise to a powerful neutron source.

The spallation target is surrounded by a fast or thermal subcritical assembly, which contains materials to be transmuted. Transport of the proton beam into the assembly is a major problem and interface devices, beam windows, etc., will have to be developed and tested. The target coolant will become strongly radioactive due to contamination with isotopes not usually encountered in conventional reactors.

### 6.4.2. Subcritical core

A subcritical core can be very similar in principle to a critical core except that the effective neutron multiplication factor is less than unity. A subcritical core cooled by liquid metal can fully utilize existing LMFR technologies.

Subcritical operation provides great freedom in design and operation. The subcritical core configuration of the ADS would allow burnup of some atypical actinide mixtures for long irradiation periods; with this kind of system it is not necessary to reach some reactivity excess in order to start the operation, because the external source maintains the steady state. To achieve the highest transmutation–incineration rates the fuel in these transmuter–burner reactors should ideally consist of pure MAs plus varying amounts of plutonium.

Criticality in a conventional reactor imposes tight constraints on the fuel specifications and cycle length. ADSs can accept fuels that would be impossible or difficult to use in critical reactors, and can extend their cycle length if necessary.

Trips and fluctuations of the incident proton beam will have to be greatly reduced, because they cause thermal shocks in the core components.

The problems that have an impact on waste management are very similar for ADSs and LMFRs if the same cooling liquid is used. Much experience has been gathered in the past on sodium as an LMFR coolant, but it is not obvious that sodium will be selected as an ADS coolant.

A lead–bismuth eutectic mixture is preferred for its thermal, neutronic and safety features, but is less attractive from the corrosion point of view.

An accelerator driven subcritical system has clear safety advantages in severe reactivity accidents. It can cope with fast ramp rate accidents, which could occur too rapidly for scram systems in critical reactors. A margin to accommodate fast reactivity insertions is important to avoid supercriticality accidents.

The consequences of cooling failure for an ADS are similar to a critical FR. A reliable beam shut-off system is therefore required for an ADS, just as a reliable scram system is required for a critical reactor. A reliable emergency decay heat removal system is required for both. A comprehensive overview of ADS technology and its comparison with FR technology has been published recently [9].

Transmutation in an ADS results, according to calculations [80], in a cleaner transmutation (i.e. with very reduced formation of higher actinides). However, the irradiation time is much longer, since the fission cross-sections are very small. At a fluence of $1.74 \times 10^{25}$ n/cm$^2$ (800 EFPD) in the conceptual MYRRHA ADS facility the final isotopic composition is as follows: 70.1% residual $^{241,242,243}$Am, 10.7% $^{238}$Pu and 10% fission products. A résumé of the computed data is given in Tables 3 and 4 and in Figs 23 and 24.

## 6.5. TRANSMUTATION ISSUES OF LONG LIVED FISSION PRODUCTS

Transmutation of LLFPs is a very difficult task, because the neutron capture cross-sections to transmute the radionuclides into short lived or stable nuclides are very small. Moreover, each neutron absorption is a net neutron loss without a compensating fission. This way, very long irradiation periods are necessary to obtain a significant depletion. Dedicated reactors with high thermal neutron fluxes and/or dedicated accelerator driven transmutation facilities (e.g. using resonance neutron absorption) are the only possible options for carrying out this very expensive endeavour.

In the short term, after discharge of the fuel, the main fission products determining the thermal load and the overall radioactivity of HLW are $^{137}$Cs and $^{90}$Sr. These two isotopes, with half-lives of 28–30 years, are not considered in P&T operations since their radioactive life is limited to about 300 years.

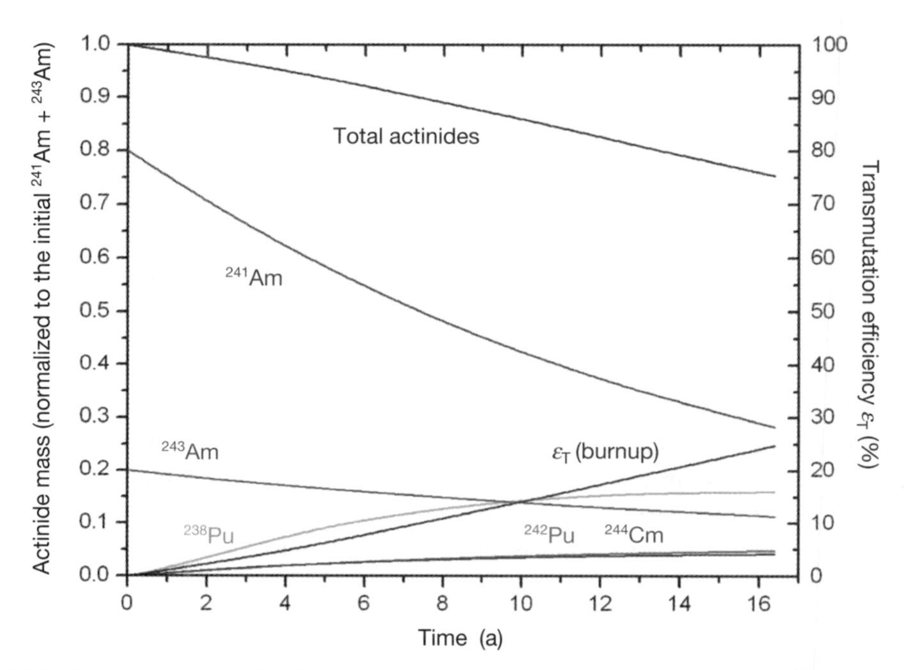


*FIG. 22. Composition of an irradiated americium target and transmutation efficiency as a function of fluence in an FR [9].*

The fission products that play an important role in the long term dose to humans originating from the back end of the fuel cycle are, in order of radiological importance, $^{129}I$, $^{99}Tc$, $^{135}Cs$, $^{93}Zr$, $^{95}Se$ and $^{126}Sn$. The associated risk varies according to the type of repository host formation.

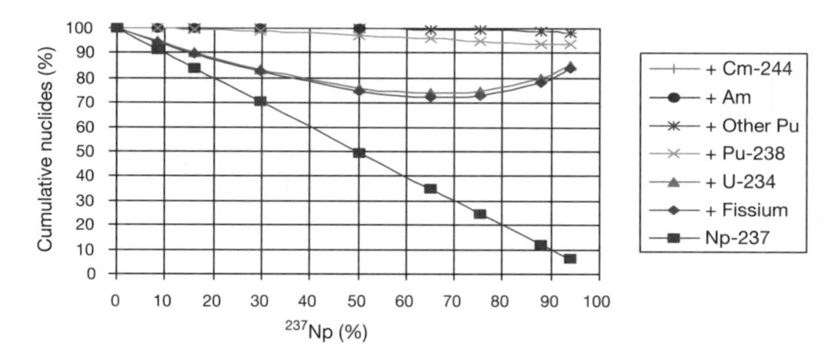


*FIG. 23. Transmutation rate of a neptunium target in a MYRRHA type ADS [9].*

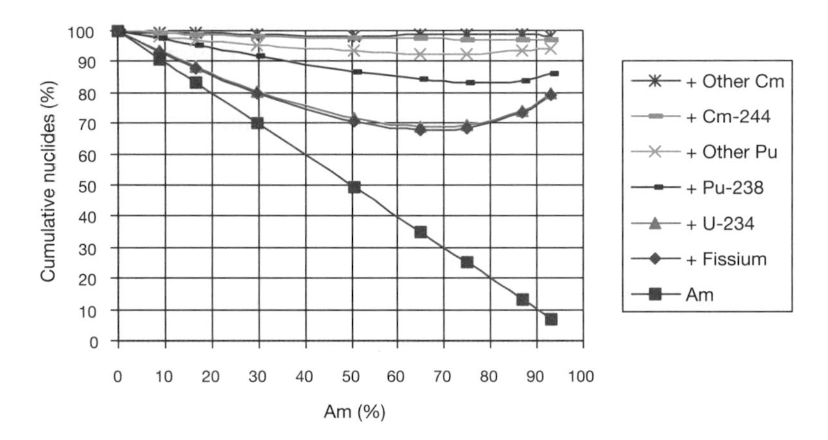


*FIG. 24. Transmutation rate of an americium target in a MYRRHA type ADS [9].*

Some activation products are also of importance in the determination of dose to humans (i.e. $^{14}C$ and $^{36}Cl$).

### 6.5.1. Iodine-129

Transmutation of $^{129}I$ is a difficult task because it has a moderate to small thermal cross-section (27 barns). Up to now, no thermally stable iodine matrix has been found, and most of the calculations have been performed for NaI, $CaI_2$ and $CeI_3$. Each of these candidates has its limitations, but $CeI_3$ seems the most promising. Two opposite safety requirements have to be fulfilled: on the one hand the confinement of the iodine compound in the target capsule during irradiation and on the other hand the discharge of the produced xenon. A vented capsule with an iodine filter is to be investigated. The upscaling of such complex irradiation procedures to industrial quantities is not obvious.

### 6.5.2. Technetium-99

Technetium-99 is one of the most important LLFPs that occur in spent fuel and in several waste streams from fuel reprocessing. Due to its long half-life (213 000 years) and the diverse chemical forms in which it can occur, its radiological significance is important if the repository surroundings are slightly oxidative. In reducing conditions of deep aquifers, it is remarkably stable and insoluble as technetium metal or $TcO_2$.

If $^{99}Tc$ is a real radiological hazard in some repository conditions, new separation technologies need to be developed. Transmutation of $^{99}Tc$ is possible if it is present as a metallic target, since the transmutation product is inactive $^{100}$Ru. Taking into account the very small thermal cross-section of 20 barns it is important to have a high thermal neutron flux, a high loading in the reactor and an optimized moderator to target radius.

### 6.5.3. Caesium-135

Caesium occurs in several isotopic forms: $^{137}$Cs, $^{134}$Cs, $^{133}$Cs and $^{135}$Cs. In terms of radiological significance $^{137}$Cs is the major constituent of HLW. Caesium-135 has a very long half-life (2 million years), but its concentration is a million times lower. Caesium is very mobile in the geosphere if not conditioned into a suitable matrix, for example glass. At present no $^{135}$Cs separation is envisaged since isotopic separation from the highly active $^{137}$Cs would be necessary in order to isolate this radionuclide. Transmutation to stable $^{136}$Cs in order to deplete $^{135}$Cs is very difficult, since stable $^{133}$Cs and $^{134}$Cs are present in the fission product mix and would generate new $^{135}$Cs during long term irradiations.

### 6.5.4. Zirconium-93

Zirconium-93 is to a certain extent similar to $^{135}$Cs, since it has a very long half-life (1.5 million years) but is present as a relatively small fraction (14%) of the total zirconium load present in the fission product mix. Separation of $^{93}$Zr involves the development and operation of isotopic separation procedures. In the longer term, it could be used in IMFs (remote fabrication).

### 6.5.5. Tin-126

Tin-126 has a half-life of 100 000 years, is partly soluble in HLLW from aqueous reprocessing but occurs also as an insoluble residue (similar to technetium). Its isolation involves a special treatment of the HLLW and the use of isotopic separation techniques. Recently, it has been proposed to add this isotope to lead spallation targets.

### 6.5.6. Carbon-14

The transmutation of $^{14}$C has not yet been considered in the P&T context. Theoretically, the $^{14}$C released from the spent fuel could partly (about 50%) be recovered from reprocessing off-gases. There is, however, not enough knowledge about the chemistry of $^{14}$C in dissolver conditions to improve this figure. Once transformed into a solid target, for example barium carbonate

($BaCO_3$), it could be stored for an infinite period. The cross-section of $^{14}C$ for thermal neutrons is nearly zero. Transmutation by charged particles in high energy accelerators is a theoretical alternative in some cases, but the practical feasibility and economy of such approaches are questionable.

Phototransmutation at 5–6 MeV can theoretically be carried out on caesium and strontium radionuclides in HLW. This technology is still at the fundamental level and will need to be further investigated [85, 86].

# 7. ADDITIONAL WASTE MANAGEMENT CONSIDERATIONS

If partitioning were to be introduced on a worldwide or continental basis the foremost impacts on waste management would be the generation of a wide variety of nuclear material and waste types that will need to wait to be either transmuted or disposed of in specific repositories. In both cases the interim engineered storage capacity will have to be increased dramatically, and the nature of the materials and their conditioning matrices may be very different from the present vitrified waste forms or conditioned spent fuel types.

However, the production of actinide free HLW is undoubtedly one of the most attractive aspects of partitioning.

An evolutionary strategy during which new waste types will occur gradually and call for the introduction of specific interim storage or disposal technologies is being considered. The following evolution might be expected:

(a) Improved reprocessing, with an impact on plutonium, neptunium, technetium and iodine management. This requires only improvement in existing reprocessing facilities.
(b) Advanced processing, with an impact on americium and curium, neptunium and plutonium residues and caesium. This requires new separation facilities.
(c) Pyrochemical processing, with the generation of TRU concentrates, technetium and iodine fractions. This requires new process developments and new separation facilities.

## 7.1. IMPACT OF IMPROVED REPROCESSING

### 7.1.1. Residual plutonium

The most obvious improvement of current aqueous reprocessing is the increase of the plutonium separation yield. Current process practice produces HLLW with 0.2% of the initial plutonium charge. By further improving the Purex process (reduced contact time, increased recycling of TBP) it may be expected that a reduction down to 0.1% plutonium contamination is within reach. With an initial load of 12 kg Pu/t HM this would reduce the plutonium concentration in vitrified HLW to 12 g Pu/t HM equivalent (~400 kg glass/t HM). On a 100 GW(e)·a scale it would reduce plutonium disposal to about 26 kg of plutonium in the 880 t mass of borosilicate glass that would be discharged annually from an all-reprocessing strategy. Further reduction of the plutonium level in vitrified HLW is not justified in comparison with the radiotoxicity of the MAs.

### 7.1.2. Neptunium

Depending on the burnup and the initial $^{235}U$ enrichment, the total neptunium inventory of 1.5–1.6 t/100 GW(e)·a is at present embedded in 880 t of borosilicate glass. If the separation technology of neptunium becomes operational, for example after a dedicated refurbishment of the first Purex extraction cycle, it could be separated for specific conditioning. If a separation yield of 99% were achieved the residual neptunium loading of the vitrified waste would be reduced to 15 kg of neptunium within the bulk of glass. This is, as discussed in the previous paragraph, of the same order of magnitude as that of the residual plutonium content in vitrified HLW.

### 7.1.3. Iodine

According to the present status of reprocessing technology, about 95–98% of iodine ($^{127}I$ and $^{129}I$) is eliminated from the dissolver solution as gaseous effluent and sent to medium level waste discharge. The capture of the elemental iodine fraction with a 99.9% yield and its transformation into an insoluble matrix would provide an additional ecological benefit. A total of 7.1 kg of iodine as fission product is produced per GW(e)·a. At a total annual discharge of 710 kg/100 GW(e), iodine waste has to be treated to reduce its dispersion in the geosphere and hydrosphere. Transformation of this molecular source term into an insoluble compound increases the mass to be handled. By transforming it into the very insoluble (but expensive) AgI or the relatively stable $Pb(IO_3)_2$ compounds, the mass increases by a factor of 1.85 and 4.37, respectively. Waste management for intermediate storage has to be addressed due to an accumulation of 1–3 t of iodine concentrate per year. Only ~82% of the iodine is made up of the very long lived $^{129}I$, but isotopic separation is not possible in this context. As already mentioned, an intermediate or retrievable disposal in a salt dome could provide very effective protection against dispersion in the geosphere.

If transmutation of iodine has to be considered, other molecular forms, for example $CeI_3$, $CaI_2$ or even untreated NaI, could be used. The problems associated with this strategy are discussed in Section 6.5.

#### 7.1.4. Technetium

Improved or advanced reprocessing could provide a small benefit for technetium management. As previously discussed, a fraction of technetium (60–80%) is present as a $TcO_4^-$ ion that is soluble in acid and is co-extracted with actinides. By scrubbing the loaded TBP stream with concentrated nitric acid, $TcO_4^-$ is washed out and separated from the actinide stream. Any action on the soluble technetium improves the plutonium partitioning in the Purex process and opens a new route to reduce the overall risk. Once separated from the process stream it could be transformed into metallic technetium and added to the insoluble fraction. Whichever strategy is chosen (storage, disposal or transmutation), transformation of all technetium forms into metallic technetium is advantageous. The total quantity of separated technetium to be managed is relatively high, and amounts to ~2 t generated annually in the discharged spent fuel. Separation would require a storage capacity of about 100 t to accept the output of a 100 GW(e)·a reactor fleet over 40 years.

Storage of separated technetium metal is feasible but constitutes a potential hazard if kept in surface facilities for a long period. In the absence of a definite choice of preferred strategy, transmutation or disposal, a special facility could be erected near the reprocessing plant for intermediate storage until the decision regarding transmutation has been taken. If the economic outlook for this technology appears too weak, transfer to a deep repository with specific geochemical characteristics compatible with a metallic matrix is the preferred option.

### 7.2. IMPACT OF ADVANCED PROCESSING

Separation of americium and curium is discussed in detail in Section 5.1.3.5 The separated quantities to be managed depend strongly on the burnup. In the present high burnup situation (~50–60 GW·d/t HM) the total quantity of americium and curium approaches 18 kg/GW(e)·a (about 64% $^{241}$Am, 26% $^{243}$Am and 10% curium). In a continental strategy of 100 GW(e)·a, a mass of 1.8 t of americium and curium would be processed annually and transferred to a dedicated storage facility.

The treatment and storage of such highly active radionuclide concentrate is a real challenge because of the $^{244}$Cm concentration in the mixture. The total decay heat of the annual output amounts to 480 kW(th) and has to be fractionated in small batches and stored under cooling. Storage as oxide is the safest approach. Storage over 100 years significantly reduces the heat dissipation (by a factor of 3.8), but it is transformed by alpha decay into the very long lived $^{239}$Pu and $^{240}$Pu isotopes. In a transmutation strategy it would be preferable to wait for several decades before starting to transform such a hot mixture into an irradiation target or MA fuel. In the subsequent target and fuel fabrication steps the neutron emission is the main problem.

All the other radionuclides (fission and activation products), as previously discussed, need to be separated isotopically and are unsuitable for further separate waste management. The same applies to $^{137}$Cs and $^{90}$Sr, which are the main contributors to radiotoxicity and are safely embedded in borosilicate glass.

### 7.2.1. Pyrochemical processing

It is too early to quantitatively predict the impact of pyrochemical processing on the interim storage of the different waste streams that will evolve from this option. Only some general but preliminary conclusions can be drawn from the proposed pyrochemical flowsheets.

(a) The main difference with partitioning operations, from the fuel cycle and waste management points of view, is the presence of the full plutonium load in these operations. The inventory of TRUs in pyrochemical processing is eight to ten times larger than the MA inventory. The facilities involved must be designed to handle this large throughput: 274 kg TRU/GW(e)·a instead of 34 kg MA/GW(e)·a. However, the actual size of the facilities per unit weight of TRU is smaller with pyrochemical processing than with aqueous processing because there are fewer criticality constraints.

(b) Transformation of oxide fuel into metallic fuel is one of the key processes, since it produces products and waste streams in a metallic form, which is different from the existing Purex waste. In order to proceed with the reduction process large quantities of lithium metal need to be injected into the reducing furnace. Residual waste streams containing partially reduced and strongly alpha contaminated Li–$Li_2O$ residues need to be processed and conditioned.

(c) Electrorefining of metallic fuel forms is the main process step in any pyrochemical process. The electrode for TRU separation is made of liquid cadmium operating at a high temperature. Most of the cadmium has to be recycled, otherwise bismuth will need to be transformed into a stable metallic waste form. The fate of these very toxic chemicals strongly contaminated with TRUs is a serious matter that needs further clarification. The final waste form is provisionally a zirconium–iron metal alloy whose stability in geohydrologic conditions has to be studied.

(d) Molten LiCl–KCl and mixed LiCl–KCl/LiF–KF are to be envisaged as molten salt media in electrorefining and TRU–fission product separation. It has to be treated at regular intervals to keep these inorganic solvents clean for electrorefining and molten salt–molten metal counter-current extraction. The final waste forms proposed for these materials are sodalite bonded salt for LiCl–KCl and apatite type zeolite for mixed chloride–fluoride mixtures. The loaded zeolites are embedded in borosilicate glass.

(e) Equipment for metal–salt separation operating at a high temperature will require frequent replacement, due to corrosion. The materials used for these vessels are very different from the conventional stainless steel types used in liquid extraction. Conditioning of these technological wastes needs further investigation. Pyrographite is at present the only candidate for electrochemical deposition of $UO_2$ and MOX in a NaCl–2CsCl–$UO_2Cl_2$ system operating at 650°C.

(f) One of the major operating conditions of pyrochemical processes involves the use of inert gas alpha tight hot cells. In principle, this severe technological condition does not have a direct impact on waste production, but it undoubtedly has an indirect impact. To keep the hot cells alpha tight and to avoid oxygen and moisture ingress are conflicting requirements, which makes the facility very complex and increases the susceptibility for incidents with major consequences on the waste output.

## 7.3. IMPACT OF TRANSMUTATION ON WASTE MANAGEMENT AND DISPOSAL

Once separation–conditioning and target fabrication processes have been developed and demonstrated on a pilot scale, transmutation by irradiation in dedicated facilities can theoretically be envisaged. Since this implies also the simultaneous development of dedicated irradiation facilities (FRs or ADSs), it will take several decades before this option could become available. However, the impact of the transmutation option starts as soon as the targets and/or fuels are ready for irradiation, which may be much earlier than the implementation of dedicated irradiation facilities. The safety and security of this option requires much attention.

(a) Once through transmutation of neptunium targets in PWRs results in a limited depletion of the initial content, by roughly 40–45%, and an overwhelming production of $^{238}Pu$ and some higher plutonium isotopes. Depending on the irradiation period (800–2460 EFPD), only limited (<18%) fissioning occurs. Transmutation in PWRs makes separated neptunium less accessible from the safeguards point of view and unusable in a nuclear explosive device. From the radiotoxic point of view this operation drastically increases ($10^3$–$10^4$) the radiotoxicity ($^{238}Pu$) and generates (~8%) a long lived source term made of $^{239,240,241,242}Pu$. The net reduction of the neptunium inventory depends on the total neptunium target depletion in the dedicated reactor and on the generation of neptunium in the uranium bearing fuel. An LWR MOX fuelled irradiation reactor is preferable to an LWR $UO_2$ core. The irradiated targets need to be stored together with the spent MOX fuel.

(b) Multiple recycling transmutation of neptunium targets can substantially deplete the initial load, at the cost of a high neutron consumption and a correspondingly higher enrichment. Separate processing facilities for the partially depleted targets are necessary to limit the $^{238}Pu$ concentration in the main stream of the reprocessing plant. About three cycles are required to reduce the inventory by ~95%. The separated fission products follow the vitrification route. The quantitative depletion of the $^{237}Np$ inventory by transmutation reduces considerably the very long term dose to a human. However, in the short and medium term (up to 800 years) the transmutation increases the radioactive burden and the heat load of the waste. Increasing the heat load has a direct effect on the engineered storage technology, on the prolongation of the intermediate storage time and on the size of the disposal facility. During these recycling operations secondary alpha waste is produced, which also has to be disposed of.

(c) A fast neutron spectrum leads to a higher yield of fission products and a somewhat lower $^{238}Pu$ and higher actinide formation, which is favourable from a radiotoxic point of view. The neutron economy (number of neutrons per fission) is much better in an FR or ADS than in a PWR, since the transmutation chain to achieve fission is much shorter. In other words, transmutation in a fast neutron spectrum is cleaner and does not generate additional long lived actinides. However, the irradiation time to reach the same level of depletion is much longer.

(d) Transmutation of separated neptunium makes its handling and storage operations in engineered storage facilities more difficult. It increases the heat load of the waste and generates secondary waste. These drawbacks have to be weighed against the very long term benefit of decreasing the dose to humans in the very distant future (millions of years). However, it gives the present generation an ethical guarantee that the present nuclear legacy will not unduly burden future civilizations. By transmuting neptunium the very long term hazard will be decreased, but it still has to be compared with the natural actinide decay chains of uranium (depleted and reprocessed).

(e) Once through transmutation of americium targets in an LWR MOX fuelled reactor results in a similar composition pattern with respect to plutonium radionuclides (30% $^{238}$Pu and ~10% higher plutonium isotopes) but shows a much higher $^{244}$Cm generation (~8%) from $^{243}$Am. The disappearance rate of $^{241}$Am is very high: about 80% in one single irradiation cycle. Since this radionuclide is the parent isotope of neptunium, transmutation in an LWR MOX reactor decreases significantly the very long term radiotoxic inventory. However, in the short and medium term the transmutation leads to a very hot target, which decays with the half-life of $^{244}$Cm (i.e. 18 years). Prolonged storage of the irradiated americium targets has to be envisaged and the radiotoxic burden will be determined by the generated plutonium isotopes. Disposal of such targets will require special conditioning techniques with a high heat removal capacity. Transmutation of americium and curium targets does not improve their safeguarding quality, since this is intrinsically a strongly irradiating mixture that cannot be manipulated without effective precautions.

(f) Multiple recycling of americium targets in LWRs is unlikely because the irradiated target is too hot for aqueous reprocessing and recycling. The presence of increased quantities of $^{244}$Cm is a major obstacle for this option. Pyrochemical multiple recycling of irradiated americium targets is a theoretical alternative that could be considered for a TRU mixture. Secondary waste generation would be a serious issue because of the increasing curium content of the mixture after each recycle.

(g) Once through irradiation of americium in a moderated subassembly of an FR has been considered as a transmutation possibility that would not need further reprocessing. Elimination of 98% of americium would require a very long irradiation period (20 years) and leads to a post-irradiative mixture of 80% fission products and ~20% actinides, primarily plutonium and curium. From a waste management point of view this option reduces the americium inventory on an atomic basis by a factor of five, but the $^{241}$Am (half-life of 432 years) is partially substituted by $^{239}$Pu (half-life of 24 400 years) and the long lived $^{243}$Am by $^{244}$Cm. The neutron economic advantage is preserved because of the fast reactor neutron economy, but the feasibility of operating a kind of EFR power reactor with many $^{241,242,243}$Am targets (2500 kg) needs further safety studies. The post-irradiation storage and disposal would be very similar to very high burnup FR MOX fuel.

(h) Multiple recycling of plutonium and MAs (TRUs) in FRs is theoretically possible. Irradiating TRUs in FRs has the advantage of reducing the inventory in proportion to the achieved burnup, independent of the TRU composition. A burnup of 18 at.% requires multiple recycling over at least five cycles to reach equilibrium between the in and out flux of TRUs. The recycling operations would be carried out in small pyrochemical units associated with an IFR reactor fleet. As fissioning would prevail on capture, a smaller amount of higher actinides would be produced and the overall mean half-life of the residual TRU inventory would decrease. Other waste forms would be introduced: an iron–zirconium matrix for metallic (fission and fissile) radionuclides together with the hulls and a sodalite–borosilicate glass mixture for the fission products. Very little has been published and most of the processes are still laboratory or pilot scale investigations.

(i) Multiple recycling of TRUs through ADS irradiation and pyrochemical fuel cycles is the latest of the proposed options. It implies the construction of a large ADS capacity (about ten times higher than for MAs). The advantage of this conceptual system is the relatively small size of the facilities required compared with those needed for aqueous reprocessing, the resistance against criticality risks and the proliferation resistance of the TRU mixture. These advantages are balanced by a lack of technical maturity of the irradiation system at the industrial scale and of processing facilities. The very high decay heat load of the treated spent fuel (192–455 kW/t HM) is compensated for by the small throughput, but the industrial feasibility of the process will have to be demonstrated. Much R&D will be required to make the process comparable with FR technology.

# 8. CONCLUSIONS

(a) The application of P&T would, if fully implemented, result in a significant decrease in the transuranic inventory to be disposed of in geologic repositories. Currently, it is believed that the inventory and radiotoxicity can be reduced by a factor of 100 to 200 and that the time scale required for the radiotoxicity to reach reference levels (natural uranium) will be reduced from over 100 000 years to between 1000 and 5000 years. To achieve these results it is believed that it would be necessary for plutonium and neptunium to be multiple recycled and for americium (curium) to be incinerated in a single deep burn step.

(b) It is unlikely that all nuclear States would have their own reprocessing and partitioning facilities to perform P&T. Similarly, disposal centres with specific technical characteristics (for individual waste types such as hot targets, iodine, etc.) would not be available in all States. Advantages may therefore be accrued by the coordination of national responsibilities to determine the long term fate of these nuclear materials. Long term storage followed by transmutation will transform some of these materials into waste, which will have to be disposed of. National and international responsibilities will need to be further defined, and international organizations can play a role in this regard.

(c) Currently, P&T is at the R&D stage. Although advanced aqueous processing has been demonstrated at the laboratory level, it needs to be scaled up to the industrial level. Pyroprocessing is still very much at the R&D stage. Before transmutation can be introduced on an industrial scale, new fuels or targets will have to be developed that may contain substantial amounts of MAs and be able to withstand high levels of irradiation. The expected time scale for such developments could be of the same order as the development of a new large commercial processing plant:
   - (i) Partitioning: aqueous: 5–10 years; pyrochemical: 10–15 years.
   - (ii) Fuel and target fabrication: 10–15 years.
   - (iii) FR reintroduction: 20–25 years.
   - (iv) ADS development: 25–40 years.

(d) Partitioning followed by conditioning (the P&C strategy) is an intermediate strategy towards P&T. Conditioning of MAs and some LLFPs as future irradiation targets or of IMF for long term storage is a dual option that could be pursued until transmutation facilities become available. The expected time scale for the start of a P&C strategy is 5–10 years if partitioning is performed by aqueous methods and 10–15 years if pyrochemical techniques are to be used.

(e) The proliferation risk potentially associated with P&C and P&T development and application, and their impact on the non-proliferation regime and on IAEA safeguards, should be addressed at the early stages of P&C and/or P&T programmes.

(f) International organizations could play an important role in monitoring transfers and inventories of separated neptunium and americium resulting from P&C and P&T projects. International organizations could also play a role in developing proliferation resistance principles and guidelines and in coordinating strategic planning and assessing the technological options for P&C and P&T systems if so requested.

(g) This study on P&T calls for more integration of the effort of the P&T and waste management communities in order to reach a common understanding of the long term issues associated with P&T and the associated radioactive waste.

# 9. RECOMMENDATIONS TO DECISION MAKERS

(a) P&T is an alternative waste management strategy that aims to reduce the very long term radiological burden of nuclear energy. It relies on the nearly quantitative recycling of long lived and highly radiotoxic nuclides. It is therefore very important to stress the crucial role of reprocessing technologies in any further P&T development. For P&T to be a viable option it will be necessary for reprocessing expertise to be preserved either in existing industrial plants or by keeping R&D projects in this field alive.

(b) Separation and recycling of plutonium is already industry practice. Partitioning of MAs by aqueous methods has been demonstrated on the laboratory and pilot scale. Improvements are expected in order to realize industrial projects that would significantly reduce the TRU content of vitrified HLW. This approach to waste management might improve the safety perception of nuclear waste disposal.

(c) Partitioning of MAs followed by waste conditioning into a very stable matrix is in the near future the most appropriate technology that would improve current waste disposal options and reduce the long term risk without affecting the transmutation option if it became practicable.

However, there are shorter term risks from using P&T technology that need to the considered.

(d) In the framework of waste volume reduction, the separation of TRUs from spent nuclear fuel by pyrochemical techniques is a new technology that needs further development in order to show its promise for the comprehensive recycling and burning of TRUs within an integrated fuel cycle fed by LWRs and FRs. This development is a long term effort that would be most realistic in the context of the revival and worldwide expansion of nuclear energy.

(e) Recycling of plutonium in LWR MOX reactors is an intermediate strategy to reduce separated plutonium stocks and to partially use its fissile content. However, full recycling and burning of plutonium is only possible when FRs become operational on the industrial scale.

(f) Since transmutation of MAs cannot yet be envisaged on the industrial scale, it would be prudent to continue this option at the research level in order to keep the possibility of the future development of dedicated transmutation reactor types open. Transmutation of TRUs needs the development of fast neutron reactors with a thermalized blanket for transmutation purposes. Fundamental research on transmutation might perhaps open new horizons for the role of transmutation in waste management.

## REFERENCES


[1] BAETSLE, L.H., et al., “Partitioning and transmutation in a strategic perspective”, Euradwaste ’99: Radioactive Waste Management Strategies and Issues (Proc. Conf. Luxembourg, 1999), Rep. EUR-19143, Office for Official Publications of the European Communities, Luxembourg (2000) 138–150.

[2] OECD NUCLEAR ENERGY AGENCY, Actinide and Fission Product Partitioning and Transmutation: Status and Assessment Report, OECD/NEA, Paris (1999).

[3] OECD NUCLEAR ENERGY AGENCY, Accelerator-driven Systems (ADS) and Fast Reactors (FR) in Advanced Nuclear Fuel Cycles: A Comparative Study, OECD/NEA, Paris (2002).

[4] US DEPARTMENT OF ENERGY, Report to Congress on Advanced Fuel Cycle Initiative: The Future Path for Advanced Spent Fuel Treatment and Transmutation Research, USDOE, Washington, DC (2003).

[5] EUROPEAN TECHNICAL WORKING GROUP ON ADS, A European Roadmap for Developing Accelerator Driven Systems (ADS) for Nuclear Waste Incineration, ENEA, Rome (2001).

[6] MAGILL, J., et al., Impact limits of partitioning and transmutation scenarios on the radiotoxicity of actinides in radioactive waste, Nucl. Energy **42** (2003) 263–277.

[7] MAGILL, J., Nuclides.net: An integrated Environment for Computations on Radionuclides and their Radiation, Springer Verlag, Heidelberg (2003).

[8] ADAMOV, E.O. (Ed.), "Radiation equivalence of radioactive waste and feed materials", White Book of Nuclear Power, Minatom, Russian Federation (2001) Ch. 13.

[9] BERTHOU, V., MAGILL, J., Transmutation Diagrams: A Tool for Assessing P&T Strategies, Rep. JRC-ITU-TN-2003/10, Institute for Transuranium Elements, Karlsruhe (2003).

[10] INOUE, T., YOKOO, T., "Advanced fuel cycle with electrochemical reduction", Global 2003—Atoms for Prosperity: Updating Eisenhower's Global Vison for Nuclear Energy (Proc. Int. Conf. New Orleans, 2003), American Nuclear Society, La Grange Park, IL (2003) CD-ROM.

[11] GERASIMOV, A.S., "Characteristics of nuclear facility required for effective transmutation of radwaste", Global 2001—Back-end of the Fuel Cycle: From Research to Solutions (Proc. Int. Conf. Paris, 2001), French Society for Nuclear Energy, Paris (2001) CD-ROM.

[12] INTERNATIONAL ATOMIC ENERGY AGENCY, The Proliferation Potential of Neptunium and Americium, GOV/1998/61, IAEA, Vienna (1998).

[13] INTERNATIONAL ATOMIC ENERGY AGENCY, The Proliferation Potential of Neptunium and Americium, GOV/1999/19/Rev.2, IAEA, Vienna (1999).

[14] MUKAIYAMA, T., et al., "Importance of double strata fuel cycle for minor actinide transmutation", 3rd Information Exchange Meeting on P&T (Proc. Mtg Cadarache, 1994), OECD/NEA, Paris (1995).

[15] BRAGIN, V., et al., "Proliferation resistance and safeguardability of innovative nuclear fuel cycles", International Safeguards Verification and Nuclear Material Security (Proc. Int. Symp. Vienna, 2001), IAEA, Vienna (2001) 322–323.

[16] KOCH, L., BETTI, M., CROMBOOM, O., MAYER, K., "Nuclear material safeguards for P and T", Global '97—International Conference on Future Nuclear Systems (Proc. Int. Conf. Yokohama, 1997), Atomic Energy Society of Japan, Tokyo (1997) 876–878.

[17] OTTMAR, H., MAYER, K., MORGENSTERN, A., ABOUSAHL, S., "Demonstration of measurement technologies for neptunium and americium verification in reprocessing", International Safeguards Verification and Nuclear Material Security (Proc. Int. Symp. Vienna, 2001), IAEA, Vienna (2001) 307–308.

[18] CHANG, Y.I., The integral fast reactor, Nucl.Technol. **88** (1989) 129.

[19] MAGILL, J., PEERANI, P., SCHENKEL, R., VAN GEEL, J., "Accelerators and (non-) proliferation", Nuclear Applications of Accelerator Technology (Proc. Topical Mtg Albuquerque, 1997), American Nuclear Society, La Grange Park, IL (1998) 440–446.

[20] MAGILL, J., PEERANI, P., (Non-) proliferation aspects of accelerator driven systems, J. Physique IV **9** (1999) 167–181.

[21] MAGILL, J., PEERANI, P., "Proliferation aspects of ADS", Impact of Accelerator-based Technologies on Nuclear Fission Safety, Rep. EUR 19608 EN, Office for Official Publications of the European Communities, Luxembourg (2000) 60–67.

[22] SCHENKEL, R., BETTI, M., MAYER, K., MAGILL, J., "New safeguards issues with advanced nuclear technologies", Science and Modern Technology For Safeguards (Proc. 3rd INMM/ESARDA Workshop, Tokyo, 2000).

[23] BAETSLE, L.H., DE RAEDT, C., Limitations of actinide recycle and fuel cycle consequences, Nucl. Eng. Des. **168** (1997) 203–210.

[24] COMETTO, M., Standardisation des outils de calcul pour les ADS et leur application à différents scénarios de transmutation de déchets, PhD Thesis, Swiss Federal Institute of Technology, Lausanne (2002).

[25] HAAS, D., et al., "Feasibility of the fabrication of americium targets", Actinide and Fission Product Partitioning and Transmutation (Proc. 5th Information Exchange Mtg Mol, 1998), Rep. EUR 18898 EN, OECD/NEA, Paris (1999) 197–205.

[26] KURATA, M., et al., "Fabrication of U-Pu-Zr metallic fuel containing minor actinides", Global '97—International Conference on Future Nuclear Systems (Proc. Int. Conf. Yokohama, 1997), Atomic Energy Society of Japan, Tokyo (1997) 1384–1389.

[27] SAKAMURA, Y., et al., "Studies on pyrochemical reprocessing for metallic and nitride fuels. Behaviour of transuranium elements in LiCl-KCl/liquid metal systems", Global '99—Nuclear Technology: Bridging the Millennia (Proc. Int. Conf. Jackson Hole, WY, 1999), American Nuclear Society, La Grange Park, IL (1999) CD-ROM.

[28] US DEPARTMENT OF ENERGY, A Roadmap for Developing Accelerator Transmutation of Waste (ATW) Technology: A Report to Congress, Rep. DOE/RW-0519, USDOE, Washington, DC (1999).

[29] MUKAIYAMA, T., et al., "Partitioning and transmutation program "OMEGA" at JAERI", Global 1995—Evaluation of Emerging Nuclear Fuel Cycle Systems (Proc. Int. Conf. Versailles, 1995), American Nuclear Society, La Grange Park, IL (1995) 110–117.

[30] GONZALEZ, E., EMBID-SEGURA, M., PEREZ-PARRA, A., "Transuranics transmutation on fertile and inert matrix lead-bismuth cooled ADS", Actinide and Fission Product Partitioning and Transmutation (Proc. 6th Information Exchange Mtg Madrid, 2000), OECD/NEA, Paris (2001) 207–218.

[31] GARCÍA SANZ, J.M., EMBID, M., FERNÁNDEZ, R., GONZÁLEZ, E., Estudio de KEFF para una Geometría Prototipo de ADS con Combustibles de Óxido de Uranio, Rep. DFN/TR-02/II-98, CIEMAT, Madrid (1998).

[32] GARCÍA SANZ, J.M., EMBID, M., FERNÁNDEZ, R., GONZÁLEZ, E., Estudio Neutrónico con Fuente Externa para Diversas Geometrías Iniciales de un Prototipo de ADS con Combustibles de Óxido de Uranio, Rep. DFN/TR-03/II-98, CIEMAT, Madrid (1998).

[33] EMBID, M., GONZALEZ, E., MARTÍN, M., Performance of Different Solid Nuclear Fuels Options for TRU Transmutation in Accelerator-driven Systems, Rep. DFN/TR-02/II-00, CIEMAT, Madrid (2000).

[34] PÉREZ PARRA, A., EMBID, M., GONZÁLEZ, E., Transuranics Transmutation on Partially Fertile (U-Zr) Matrix Lead-Bismuth Cooled ADS, Rep. CIEMAT/DFN/TR-01/PC-01, CIEMAT, Madrid (2001).

[35] GARCÍA SANZ, J.M., FERNÁNDEZ, R., EMBID, M., GONZÁLEZ, E.J., "Isotopic composition simulation of the sequence of discharges from a thorium TRU's, lead cooled ADS", ADDTA '99: Accelerator Driven Transmutation Techniques and Applications (Proc. 3rd Int. Conf. Prague, 1999) CD-ROM.

[36] EMBID-SEGURA, M., GONZÁLEZ, E., PÉREZ PARRA, A., "Elimination of the nuclear waste coming from the back-end of the double strata scheme in a non-equilibrium cycle", Accelerator Driven Transmutation Technology and Applications (Proc. Int. Conf. Reno, NV, 2001) CD-ROM.

[37] EWING, R.C., LUTZE, W., WEBER, W., Zircon: A host for the disposal of weapon plutonium, J. Mater. Res. **10** (1995) 243–246.

[38] RIOS, S., SALJE, E.K., ZHANG, M., EWING, R.C., Amorphization in zircon: Evidence for direct impact damage, J. Phys. Condens. Matter **12** (2000) 1–12.

[39] WEBER, W.J., EWING, R.C., MELDRUM, A., The kinetics of alpha-decay-induced amorphization in zircon, J. Nucl. Mater. **250** (1997) 147–155.

[40] YAMASHITA, T., et al., In-pile irradiation of plutonium rock-like oxide fuels, J. Nucl. Mater. **274** (1999) 98–104.

[41] DEGUELDRE, C., et al., "Relevant material properties of a zirconia based inert matrix fuel for plutonium", TopFuel 2001 (Proc. Int. Conf. Stockholm, 2001), European Nuclear Society, Brussels (2001) CD-ROM.

[42] RAISON, P.E., HAIRE, R.G., "Transmutation of americium and curium incorporated in zirconia-based host materials", Global 2001—Back-end of the Fuel Cycle: From Research to Solutions (Proc. Int. Conf. Paris, 2001), French Society for Nuclear Energy, Paris (2001) CD-ROM.

[43] HORWITZ, E.P., SCHULZ, W.W., "The TRUEX Process: A vital tool for disposal of US defense nuclear waste", New Separation Chemistry Techniques for Radioactive Waste and Other Specific Applications (Proc. Sem. Rome, 1990), Elsevier Applied Science, Barking, UK (1991) 21–29.

[44] MADIC, C., et al., "Actinide partitioning from high level waste using the Diamex process", Nuclear Fuel Reprocessing and Waste Management (Proc. 4th Int. Conf. London, 1994), Rep. CEA-CONF-12297, Commissariat à l'énergie atomique, Marcoule (1994).

[45] MORITA, Y., et al., Actinide partitioning from HLW in a continuous DIDPA extraction process by means of centrifugal extractor, Solvent Extr. Ion Exch. **14** (1996) 385–400.

[46] CHONGLI, Song, JIANCHEN, Wang, JUNFU, Lian, "Treatment of high saline HLLW by total partitioning process", Global '97—International Conference on Future Nuclear Systems (Proc. Int. Conf. Yokohama, 1997), Atomic Energy Society of Japan, Tokyo (1997) 475–480.

[47] MURALI, M.S., et al., Use of a mixture of TRPO and TBP for the partitioning of actinides from high-level waste of Purex origin, Solvent Extr. Ion Exch. **19** (2001) 61–77.
[48] DEL CUL, C.D., et al., Citrate-based "TALSPEAK" Lanthanide-Actinide Separation Process, Rep. ORNL/TM-12784, Oak Ridge National Lab., TN (1994).
[49] LILJENZIN, J.O., PERSSON, G., SVANTESSON, I., WINGEFORS, S., The CTH process for HLLW treatment. Pt. 1. General description and process design, Radiochim. Acta **35** (1984) 155–162.
[50] ZHU, Yongjun, CHEN, Jing, JIAO, Rongzhou, "Hot test and process parameter calculation of purified CYANEX 301 extraction for separating Am and fission product lanthanides", Global '97—International Conference on Future Nuclear Systems (Proc. Int. Conf. Yokohama, 1997), Atomic Energy Society of Japan, Tokyo (1997) 581–585.
[51] MANCHANDA, V.K., Bhabha Atomic Research Centre, Mumbai, personal communication, 2003.
[52] MODOLO, G., et al., "The ALINA-process for An(III)/Ln(III) group separation from strong acidic medium", Global '99—Nuclear Technology: Bridging the Millennia (Proc. Int. Conf. Jackson Hole, WY, 1999), American Nuclear Society, La Grange Park, IL (1999) CD-ROM.
[53] KOLARIK, Z., MÜLLICH, U., "Separation of Am(III) and Eu(III) by selective solvent extraction with N-donor extractants", Global '97—International Conference on Future Nuclear Systems (Proc. Int. Conf. Yokohama, 1997), Atomic Energy Society of Japan, Tokyo (1997) 586–591.
[54] MADIC, C., et al., ""PARTNEW": A European research program (2000-2003) for partitioning of minor actinides from high level liquid wastes", Global 2001—Back-end of the Fuel Cycle: From Research to Solutions (Proc. Int. Conf. Paris, 2001), French Society for Nuclear Energy, Paris (2001) CD-ROM.
[55] MADIC, C., LECOMTE, M., BARON, P., BOULLIS, P., Separation of long-lived radionuclides from high active nuclear waste, Comptes Rendus Physique **3** (2002) 797–811.
[56] GLATZ, J.P., et al., "Demonstration of partitioning schemes proposed in the frame of P&T studies using genuine fuel", Global '99—Nuclear Technology: Bridging the Millennia (Proc. Int. Conf. Jackson Hole, WY, 1999), American Nuclear Society, La Grange Park, IL (1999) CD-ROM.
[57] STEVENS, C.E., The EBRII Fuel Cycle Story, American Nuclear Society, La Grange Park, IL (1987).
[58] BYCHKOV, A.V., et al., "Pyroelectrochemical reprocessing of irradiated FBR-MOX fuel. 3. Experiment on high burn-up fuel for the BOR-60 reactor", Global '97—International Conference on Future Nuclear Systems (Proc. Int. Conf. Yokohama, 1997), Atomic Energy Society of Japan, Tokyo (1997) 912–917.
[59] ATW SEPARATIONS TECHNOLOGIES AND WASTE FORMS TECHNICAL WORKING GROUP, A Roadmap for Developing ATW Technology: Separations & Waste Forms Technology, Rep. ANL-99/15, Argonne National Laboratory, Argonne, IL (1999).

[60] INSTITUTE FOR TRANSURANIUM ELEMENTS, ITU Activity Report 1999, Rep. EUR 19054, Institute for Transuranium Elements, Karlsruhe (1999) 34–45.

[61] KOYAMA, T., et al., "Demonstration of pyrometallurgical processing for metal fuel and HLW", Global 2001—Back-end of the Fuel Cycle: From Research to Solutions (Proc. Int. Conf. Paris, 2001), French Society for Nuclear Energy, Paris (2001) CD-ROM.

[62] KOYAMA, T., et al., Study on molten salt electrorefining of U-Pu-Zr alloy, J. Nucl. Sci. Technol., Suppl. 3 (2002) 765–768.

[63] BYCHKOV, A.V., et al., "Pyroelectrochemical reprocessing of spent FBR fuel V. Testing and demonstration of $UO_2$ and MOX flowsheets on the real spent fuel of BOR-60 reactor", Global 2001—Back-end of the Fuel Cycle: From Research to Solutions (Proc. Int. Conf. Paris, 2001), French Society for Nuclear Energy, Paris (2001) CD-ROM.

[64] INOUE, T., TANAKA, H., "Recycling of actinides produced in LWR and FBR fuel cycle by applying pyrometallurgical process", Global '97—International Conference on Future Nuclear Systems (Proc. Int. Conf. Yokohama, 1997), Atomic Energy Society of Japan, Tokyo (1997) 646–652.

[65] ARAI, Y., IWAI, T., NAKAJIMA, K., SUZUKI, Y., "Recent progress of nitride fuel development in JAERI. Fuel property, irradiation behaviour and application to dry processing", Global '97—International Conference on Future Nuclear Systems (Proc. Int. Conf. Yokohama, 1997), Atomic Energy Society of Japan, Tokyo (1997) 664–669.

[66] RENARD, A., et al., "Implications of plutonium and americium recycling on MOX fuel fabrication", Global 1995—Evaluation of Emerging Nuclear Fuel Cycle Systems (Proc. Int. Conf. Versailles, 1995), American Nuclear Society, La Grange Park, IL (1995) 1683–1690.

[67] WAKABAYASHI, T., TAKAHASHI, K., YANAGISAWA, T., Feasibility studies of plutonium and minor actinide burning in fast reactors, Nucl. Technol. **118** (1997) 14.

[68] HAAS, D., SOMERS, J., CHAROLLAIS, F., "Innovative fabrication of fuels and targets for Pu recycling and minor actinides transmutation", TopFuel '99 (Proc. Int. Topical Mtg Avignon, 1999), French Society for Nuclear Energy, Paris (1999) 137–146.

[69] SCHRAM, R.P.C., et al., "Irradiation experiments and fabrication technology of inert matrix fuels for the transmutation of actinides", Global '99—Nuclear Technology: Bridging the Millennia (Proc. Int. Conf. Jackson Hole, WY, 1999), American Nuclear Society, La Grange Park, IL (1999) CD-ROM.

[70] RAISON, P.E., HAIRE, R.G., Zirconia-based materials for transmutation of americium and curium: Cubic stabilized zirconia and zirconium oxide pyrochlores, Prog. Nucl. Energy **38** (2001) 251–254.

[71] CHAWLA, R., et al., "First experimental results from neutronics and in pile testing of a Pu-Er-Zr oxide inert matrix fuel", TopFuel 2001 (Proc. Int. Conf. Stockholm, 2001), European Nuclear Society, Brussels (2001) CD-ROM.

[72] FERNANDEZ, A., et al., Qualification of an advanced fabrication process based on the infiltration of actinide solution, J. Am. Ceram. Soc. **85** (2002) 694.

[73] ARAI, Y., et al., "Experimental research on nitride fuel cycle in JAERI", Global '99—Nuclear Technology: Bridging the Millennia (Proc. Int. Conf. Jackson Hole, WY, 1999), American Nuclear Society, La Grange Park, IL (1999) CD-ROM.

[74] OAK RIDGE NATIONAL LABORATORY, Molten-salt Reactor Program, Rep. ORNL-3708 UC-80-Reactor Technology, Oak Ridge National Lab., TN (1964).

[75] BOWMAN, C.D., Accelerator-driven systems for nuclear waste transmutation, Ann. Rev. Nucl. Part Sci. **48** (1998) 505–556.

[76] DONALDSON, N., et al., "Pyrochemistry: From flowsheet to industrial facility", Global 2001—Back-end of the Fuel Cycle: From Research to Solutions (Proc. Int. Conf. Paris, 2001), French Society for Nuclear Energy, Paris (2001) CD-ROM.

[77] RABOTNOV, N., Minatom, Moscow, personal communication, 2003.

[78] BAETSLÉ, L.H., "Impact of high burnup irradiation and multiple recycle of plutonium and minor actinides", 3rd Information Exchange Meeting on P&T (Proc. Mtg Cadarache, 1994), OECD/NEA, Paris (1995).

[79] DEGUELDRE, C., Commissariat à l'énergie atomique, France, personal communication, 2003.

[80] DE RAEDT, C., et al., "Transmutation and incineration of MAs in PWRs, MTRs and ADSs", Advanced Reactors with Innovative Fuels (Proc. Conf. Chester, 2001), OECD/NEA, Paris (2002) 445–460.

[81] ROME, M., HARISLUR, A., HERNIOU, E., ZAETTA, A., TOMMASI, J., "Use of fast reactors to burn long-life actinides, especially Am, produced by current reactors", PHYSOR96: Physics of Reactors (Proc. Int. Conf. Mito, 1996), Vol. 4, Atomic Energy Society of Japan, Tokyo (1996) 52–62.

[82] LANGUILLE, A., et al., "CAPRA core studies: The oxide reference option", Global 1995—Evaluation of Emerging Nuclear Fuel Cycle Systems (Proc. Int. Conf. Versailles, 1995), American Nuclear Society, La Grange Park, IL (1995) 874–881.

[83] BAETSLÉ, L.H., DE RAEDT, C., Some aspects of risk reduction strategy by multiple recycling in fast burner reactors of the Pu and MA inventories, Nucl. Eng. Des. **172** (1997) 359–366.

[84] RUBBIA, C., et al., An Energy Amplifier for Cleaner and Inexhaustible Nuclear Energy Production Driven by a Particle Accelerator, Rep. CERN/AT/93-47(ET), European Organization for Nuclear Research, Geneva (1993).

[85] MAGILL, J., Laser transmutation of iodine-129, Appl. Phys. B **77** (2003) 387–390.

[86] SCHWOERER, H., et al., Fission of actinides using a tabletop laser, Europhys. Lett. **61** (2003) 47–52.

## Annex I

# NATURAL AND ARCHAEOLOGICAL ANALOGUES

## I–1. NATURAL ANALOGUES

There are many radioactive materials that occur naturally and that can be found in rocks, sediments, etc. [I–1]. In particular, uranium, which is the main component of nuclear fuel, occurs in nature. By studying the distribution in nature, information can be obtained on the movement of uranium in rocks and groundwater.

Natural analogues provide a way of informing the public about the principles on which repositories are built, without using complex mathematical demonstrations of 'safety' and 'risk'. One of the concepts that can be presented using analogues is the very slow degradation of materials over thousands of years. Some notable natural analogues are:

(a) The Dunarobba forest. Dead tree trunks in the Dunarobba forest in Italy are approximately 2 million years old. In contrast to typical fossilized trees, the Dunarobba trees are still composed of wood, since the wood has been preserved by surrounding clay, which stopped oxygenated waters from reaching the wood. The Dunarobba trees are of relevance in repository concepts since wood is considered to be analogous to the organic–cellulosic materials that comprise a part of the waste.

(b) The Needle's Eye. This site in southwest Scotland comprises a sea cliff in which the mineralized veins of uranium and other metals are partly exposed. The dissolution process leads to a preferential loss of $^{234}U$ relative to $^{238}U$. The site is ideal for investigating radionuclide migration.

(c) The Oklo natural fission reactor. Fission reactions occurred at Oklo in West Africa intermittently $10^5$–$10^6$ years ago. The natural fission reactors at Oklo can be considered as analogues for very old radioactive waste repositories and can be used to study the transport behaviour of transuranic radionuclides and the stability of uranium minerals that have undergone criticality.

## I–2. SOCIETAL ANALOGUES

In the event that P&T is introduced, it is likely that one can considerably relax the period of time over which waste repositories must confine the waste.

Based on the P&T goals discussed above, it may only be necessary to demonstrate containment over a period of 500–700 years rather than the hundreds of thousands of years in the case of natural decay. On this much reduced time scale, there are many examples of human-made objects that have withstood degradation. Some examples are the buildings erected by the Egyptian pharaohs and Roman emperors and many objects of daily use in sophisticated ancient cultures.

(a) The Egyptian pyramids. According to recent findings, the Egyptian pyramids were constructed using natural cement-like materials (natronite), which have withstood the natural weather conditions in the semi-desertic conditions of northern Egypt for 4700 years. Many religious objects in wood, metal and textile have been found undamaged in the tombs.

(b) Roman buildings. The most remarkable example is the Pantheon, which was constructed by the Roman emperor Vespasianus (118–125 A.D.) and still stands undamaged after 2000 years in the middle of the bustling city of Rome. The Roman 'concrete' was made by mixing volcanic ashes (puzzolanic earth) with sand and water. Modern Portland cement species can be made as resistant as these old archaeological examples if sufficient attention is given to the nature and firing temperature (~1450°C) of the hydrated calcium silicates before mixing with sand and water.

(c) The Inchtuthil Roman nails. These Roman nails were discovered in the 1950s in Perth, UK. Over one million nails had been buried in a 5 m deep pit and covered with compacted earth. The outermost nails were badly corroded. The innermost nails, however, showed only very limited corrosion. This was attributed to the fact that the outermost nails removed the oxygen from the infiltrating groundwater such that by the time the water came into contact with the innerlying nails it was less corrosive. In the same way, the large volumes of iron in waste containers are expected to maintain chemically reducing conditions in an environment that is oxygen rich due to the radiolytic decomposition of water.

(d) The Kronan cannon. One of the bronze cannons on board the Kronan, a Swedish warship sunk in 1676, remained partly buried in a vertical position, muzzle down in clay sediments, since the ship sank. This cannon is a good analogue for canisters to be used in the Swedish and Finnish spent fuel repositories, which have a copper outside shell. From an analysis of the cannon surface, a corrosion rate of 0.15 $\mu$m/a was established. At this rate of corrosion it would take some 70 000 years to corrode a 1 cm thickness of copper. This provides evidence for the very long life of copper spent fuel canisters in a repository.

(e) Hadrian's wall. Hadrian's wall in the UK was built starting in A.D. 122 from stone blocks cemented together. The wall is of interest as an analogue due to the longevity of the Roman cement used to bind the stones together. Modern Portland cement is very similar chemically and mineralogically. From these studies, conclusions can be drawn with regard to the stability and longevity of modern cement in repositories.

## REFERENCE TO ANNEX I


[I–1] MAGILL, J., et al., Impact limits of partitioning and transmutation scenarios on the radiotoxicity of actinides in radioactive waste, Nucl. Energy **42** (2003) 263–277.

## Annex II

## CURRENT STUDIES ON INERT MATRIX FUEL

The physical properties of stabilized zirconia depend on the choice of stabilizer as well as on other dopants, for example burnable poison or fissile material. As a result of an iterative study, a $(Er,Y,Pu,Zr)O_{2-x}$ solid solution with a defined fraction of fissile and burnable poison was selected [II–1–II–3]. This material has been fabricated by two different routes. The first was a wet route including coprecipitation of the oxihydroxide phase by internal gelation, starting with the nitrate solutions [II–4]. The second was a dry route using an attrition milling unit adapted to the zirconia material, starting with the powder of the constituent oxides [II–5]. The materials were then compacted and sintered under a controlled atmosphere.

These IMF materials, solid solutions or composites are currently being tested for their thermodynamic properties and behaviour under irradiation. The first irradiation tests were carried out using ion beams from accelerators. Samples were studied during irradiation and implantations were performed to further study the behaviour of selected elements in the IMF.

Irradiation tests in research reactors are ongoing, and example cases are discussed below.

JAERI, the PSI and the Nuclear Research and Consultancy Group are performing irradiation experiments on YSZ doped with erbia and plutonia and YSZ doped with gadolinia and plutonia phases embedded in spinel ($MgAl_2O_4$) in the Petten HFR [II–6]. The project is called OTTO and the samples will be irradiated for at least 22 cycles in the HFR in-core position H8. The total neutron fluence will be approximately $1 \times 10^{26}$ $m^{-2}$.

The Korea Atomic Energy Research Institute, the PSI, British Nuclear Fuels and the Organisation for Economic Co-operation and Development (OECD) are also performing $PuO_2$–$ZrO_2$ IMF irradiation in the Halden reactor [II–7]. The irradiation with IMF based on yttria stabilized zirconia and MOX fuel (three IMF and three MOX specially fabricated rodlets) is ongoing as a part of the OECD Halden reactor project. The aim of this experiment is to measure behaviour under irradiation and the safety relevant parameters of these fuels. The irradiation experiment started towards the end of June 2000 and is scheduled to last until 2005.

The principal aim of the experiment is to measure the centre line temperature and its change with burnup, fission gas release, densification, swelling and the general thermal behaviour of the fuels. The first cycle lasted 120 days, and an average assembly burnup of 47 kW·d/$cm^3$ (i.e. 4.5 MW·d/kg MOX for MOX fuel) was attained. Note that this unconventional unit for burnup (energy released per unit volume) was chosen in order to be able to compare IMF and MOX fuel with their very different densities.

Figure II–1 shows the measured fuel centre line temperatures over the course of the first cycle. After 15.5 days the maximum power of the IMF and MOX rodlets (~25 kW/m average) was achieved. The measured temperatures are within the expected range for both IMF and MOX, the higher temperatures in the IMF rodlets reflecting the significantly lower thermal conductivity of IMF. In spite of the strong densification, the IMF temperature remains relatively stable. This might be explained by the formation of a central void region, while the outer pellet diameter (and hence the gap size) is unaffected. However, only post-irradiation examinations will reveal the effective process.

It is important to point out that zirconia based IMF production may be achieved using MOX technology.

For the material qualification, relevant fuel properties were considered. Among them, good behaviour and stability under irradiation, efficient retention of fission products and extremely low solubility (e.g. about $10^{-9}$M [II–8]) are the key properties of the fuel in the reactor as well as for the geologic disposal of the spent fuel.

Many questions are still open in this new research field, for example the irradiation stability of the chosen material, the fuel behaviour under power

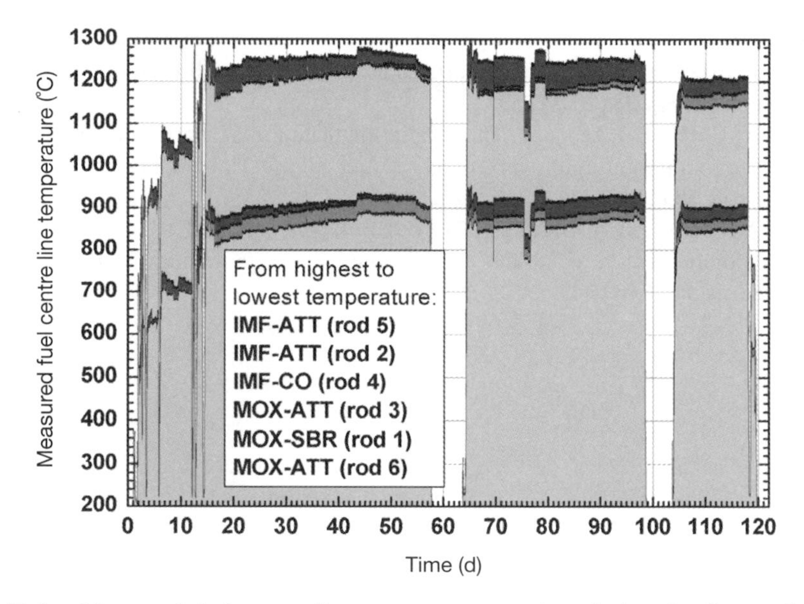


*FIG. II–1. Measured fuel centre line temperature vs. time during the first cycle. The upper profiles are the zirconia IMF. The lower curves are the MOX result [II–7].*

transients, the change in thermal conductivity with irradiation, the fission gas retention potential, the leaching behaviour of spent fuel and the several different possibilities to enhance the thermal conductivity and/or plutonium burning efficiency by modification of the additives. Although yttria stabilized zirconia is the matrix favoured by many research groups, due to recent experimental results, a change of the basic matrix material is still possible and would, if necessary, broaden the research field again.

## REFERENCES TO ANNEX II


[II–1] DEGUELDRE, C., KASEMEYER, U., BOTTA, F., LEDERGERBER, G., Plutonium incineration in LWR by a once through cycle with a rock-like fuel, Mater. Res. Soc. Symp. Proc. **412** (1996) 15.

[II–2] DEGUELDRE, C., PARATTE, J.M., Basic properties of a zirconia-based fuel material for light water reactors, Nucl. Technol. **123** (1998) 21.

[II–3] DEGUELDRE, C., et al., Behaviour of implanted xenon in yttria-stabilised zirconia as inert matrix of a nuclear fuel, J. Nucl. Mater. **289** (2001) 115.

[II–4] LEDEGERBER, G., DEGUELDRE, C., HEIMGARTNER, P., POUCHON, M., KASEMEYER, U., Inert matrix fuel for the utilisation of plutonium, Prog. Nucl. Energy **38** (2001) 301.

[II–5] LEE, Y.W., et al., Preparation of simulated inert matrix fuel with different powders by dry milling method, J. Nucl. Mater. **274** (1999) 7.

[II–6] SCHRAM, R.P., et al., Design and fabrication aspects of a plutonium-incineration experiment using inert matrices in a "once-through-then-out" mode, Prog. Nucl. Energy **38** (2001) 259.

[II–7] KASEMEYER, U., et al., The irradiation test of inert matrix fuel in comparison to uranium plutonium mixed oxide fuel at the Halden reactor, Prog. Nucl. Energy **38** (2001) 309.

[II–8] POUCHON, M., CURTI, E., DEGUELDRE, C., TOBLER, L., The influence of carbonate on the solubility of zirconia: New experimental data, Prog. Nucl. Energy **38** (2001) 443.

# ABBREVIATIONS

| | |
|---|---|
| ADS | accelerator driven system |
| ALI | annual limit on intake |
| An | actinide |
| An(III) | trivalent actinide |
| BTP | bis-triazinyl-1,2,4-pyridines |
| CSA | comprehensive safeguards agreement |
| DF | decontamination factor |
| DIAMEX | diamide extraction |
| FBuR | fast burner reactor |
| FSV | flowsheet verification |
| FR | fast reactor |
| HAR | high active raffinate |
| HAW | high active waste |
| HFR | High Flux Reactor |
| HLLW | high level liquid waste |
| HLW | high level waste |
| ICP–MS | inductively coupled plasma mass spectrometry |
| IDMS | isotope dilution mass spectrometry |
| IFR | integral fast reactor |
| IMF | inert matrix fuel |
| LILW | low and intermediate level waste |
| LLFP | long lived fission product |
| LMFBR | liquid metal fast breeder reactor |
| LMFR | liquid metal fast reactor |
| Ln | lanthanide |
| Ln(III) | trivalent lanthanide |
| LWR | light water reactor |
| MA | minor actinide |

| | |
|---|---|
| MOX | mixed oxide |
| OTC | once through fuel cycle |
| P&T | partitioning and transmutation |
| RFC | reprocessing fuel cycle |
| TBP | tri-n-butylphosphate |
| TRU | transuranic element |
| WWER | water cooled, water moderated power reactor |

# CONTRIBUTORS TO DRAFTING AND REVIEW

| | |
|---|---|
| Achuthan, P.V. | Bhabha Atomic Research Centre, India |
| Baetslé, L.H. | SCK•CEN, Belgium |
| Burcl, R. | International Atomic Energy Agency |
| Degueldre, C. | Paul Scherrer Institute, Switzerland |
| Embid-Segura, M. | CIEMAT, Spain |
| Gherardi, G. | ENEA, Italy |
| Inoue, T. | Central Research Institute of the Electric Power Industry, Japan |
| Liu, Z. | International Atomic Energy Agency |
| Madic, C. | Commissariat à l'énergie atomique, France |
| Magill, J. | Institute for Transuranium Elements/European Commission, Germany |
| Parker, D. | British Nuclear Fuels, United Kingdom |
| Rabotnov, N. | Minatom, Russian Federation |
| Sjöblom, K.-L. | International Atomic Energy Agency |
| Stewart, L. | US Department of Energy, United States of America |
| Yoo, J.-H. | Korea Atomic Energy Research Institute, Republic of Korea |

**Consultants Meetings**

Vienna, Austria: 15–19 October 2001, 7–11 April 2003

**Technical Meeting**

Vienna, Austria: 9–13 September 2002

No. 17, January 2003

# WHERE TO ORDER IAEA PUBLICATIONS

**In the following countries** IAEA publications may be purchased from the sources listed below, or from major local booksellers. Payment may be made in local currency or with UNESCO coupons.

**AUSTRALIA** Hunter Publications, 58A Gipps Street, Collingwood, Victoria 3066
Telephone: +61 3 9417 5361 • Fax: +61 3 9419 7154
*E-mail: admin@tekimaging.com.au • Web site: http://www.hunter-pubs.com.au*

**BELGIUM** Jean de Lannoy, avenue du Roi 202, B-1190 Brussels • Telephone: +32 2 538 43 08 • Fax: +32 2 538 08 41
*E-mail: jean.de.lannoy@infoboard.be • Web site: http://www.jean-de-lannoy.be*

**CANADA** Renouf Publishing Company Ltd., 1-5369 Canotek Rd., Ottawa, Ontario, K1J 9J3
Telephone: +613 745 2665 • Fax: +613 745 7660
*E-mail: order.dept@renoufbooks.com • Web site: http://www.renoufbooks.com*

**CHINA** IAEA Publications in Chinese: China Nuclear Energy Industry Corporation, Translation Section, P.O. Box 2103, Beijing

**FINLAND** Akateeminen Kirjakauppa, PL 128 (Keskuskatu 1), FIN-00101 Helsinki
Telephone: +358 9 121 4418 • Fax: +358 9 121 4435
*E-mail: sps@akateeminen.com • Web site: http://www.akateeminen.com*

**FRANCE** Nucléon Editions, 20, rue Rosenwald, F-75015 Paris
Telephone: +33 1 42 50 15 50 • +33 6 86 92 69 39 • Fax +33 1 53 68 14 93 • *E-mail: nucleon@nucleon.fr*

Form-Edit, 5, rue Janssen, P.O. Box 25, F-75921 Paris Cedex 19
Telephone: +33 1 42 01 49 49 • Fax: +33 1 42 01 90 90 • *E-mail: formedit@formedit.fr*

**GERMANY** UNO-Verlag, Vertriebs- und Verlags GmbH, Am Hofgarten 10, D-53113 Bonn
Telephone: +49 228 94 90 20 • Fax: +49 228 94 90 222
*E-mail: bestellung@uno-verlag.de • Web site: http://www.uno-verlag.de*

**HUNGARY** Librotrade Ltd., Book Import, P.O. Box 126, H-1656 Budapest
Telephone: +36 1 257 7777 • Fax: +36 1 257 7472 • *E-mail: books@librotrade.hu*

**INDIA** Allied Publishers Limited, 1-13/14, Asaf Ali Road, New Delhi 110002
Telephone: +91 11 3233002, 004 • Fax: +91 11 3235967
*E-mail: aplnd@del2.vsnl.net.in • Web site: http://www.alliedpublishers.com*

**ITALY** Libreria Scientifica Dott. Lucio di Biasio "AEIOU", Via Coronelli 6, I-20146 Milan
Telephone: +39 02 48 95 45 52 or 48 95 45 62 • Fax: +39 02 48 95 45 48

**JAPAN** Maruzen Company, Ltd., 13-6 Nihonbashi, 3 chome, Chuo-ku, Tokyo 103-0027
Telephone: +81 3 3275 8582 • Fax: +81 3 3275 9072
*E-mail: journal@maruzen.co.jp • Web site: http://www.maruzen.co.jp*

**NETHERLANDS** Martinus Nijhoff International, Koraalrood 50, P.O. Box 1853, 2700 CZ Zoetermeer
Telephone: +31 793 684 400 • Fax: +31 793 615 698 • *E-mail: info@nijhoff.nl • Web site: http://www.nijhoff.nl*

Swets and Zeitlinger b.v., P.O. Box 830, 2160 SZ Lisse
Telephone: +31 252 435 111 • Fax: +31 252 415 888 • *E-mail: infoho@swets.nl • Web site: http://www.swets.nl*

**SLOVAKIA** Alfa Press, s.r.o, Račianska 20, SQ-832 10 Bratislava • Telephone/Fax: +421 7 566 0489

**SLOVENIA** Cankarjeva Zalozba d.d., Kopitarjeva 2, SI-1512 Ljubljana
Telephone: +386 1 432 31 44 • Fax: +386 1 230 14 35
*E-mail: import.books@cankarjeva-z.si • Web site: http://www.cankarjeva-z.si/uvoz*

**SPAIN** Díaz de Santos, S.A., c/ Juan Bravo, 3A, E-28006 Madrid
Telephone: +34 91 781 94 80 • Fax: +34 91 575 55 63 • *E-mail: compras@diazdesantos.es • carmela@diazdesantos.es barcelona@diazdesantos.es • julio@diazdesantos.es • Web site: http://www.diazdesantos.es*

**UNITED KINGDOM** The Stationery Office Ltd, International Sales Agency, 51 Nine Elms Lane, London SW8 5DR
Telephone: +44 870 600 552 • Fax: +44 207 873 8416
*E-mail: Orders to: book.orders@theso.co.uk • Enquiries to: ipa.enquiries@theso.co.uk*
*Web site: http://www.the-stationery-office.co.uk*

**On-line orders**
DELTA Int. Book Wholesalers Ltd., 39 Alexandra Road, Addlestone, Surrey, KT15 2PQ
*E-mail: info@profbooks.com • Web site: http://www.profbooks.com*

*Books on the Environment*
SMI (Distribution Services) Limited, P.O. Box 119, Stevenage SG1 4TP, Hertfordshire
*E-mail: customerservices@earthprint.co.uk • Web site: http://www.earthprint.co.uk*

**UNITED STATES OF AMERICA** Bernan Associates, 4611-F Assembly Drive, Lanham, MD 20706-4391
Telephone: 1-800-274-4447 (toll-free) • Fax: (301) 459-0056/1-800-865-3450 (toll-free)
*E-mail: query@bernan.com • Web site: http://www.bernan.com*

Renouf Publishing Company Ltd., 812 Proctor Ave., Ogdensburg, NY, 13669
Telephone: +888 551 7470 (toll-free) • Fax: +888 568 8546 (toll-free)
*E-mail: order.dept@renoufbooks.com • Web site: http://www.renoufbooks.com*

**Orders and requests for information** may also be addressed directly to:

**Sales and Promotion Unit, International Atomic Energy Agency**
**Wagramer Strasse 5, P.O. Box 100, A-1400 Vienna, Austria**
**Telephone: +43 1 2600 22529 (or 22530) • Fax: +43 1 2600 29302**
***E-mail: sales.publications@iaea.org • Web site: http://www.iaea.org/worldatom/Books***

04-23811